Noise Control

Edited by Marco Caniato
and Federica Bettarello

Published in London, United Kingdom

Noise Control
http://dx.doi.org/10.5772/intechopen.96830
Edited by Marco Caniato and Federica Bettarello

Contributors
Antonio Barba, Juan M. Martinez-Orozco, José A. Ballesteros, María Jesús Ballesteros, Samuel Quintana, Marcos D. David Fernandez, Bijay Kumar Swain, Chidananda Prasad Das, Shreerup Goswami, Tetsuya Miyoshi, Alice Elizabeth González

First published in London, United Kingdom, 2023 by IntechOpen
IntechOpen is the global imprint of INTECHOPEN LIMITED, registered in England and Wales, registration number: 11086078, 5 Princes Gate Court, London, SW7 2QJ, United Kingdom

British Library Cataloguing-in-Publication Data
A catalogue record for this book is available from the British Library

Additional hard and PDF copies can be obtained from orders@intechopen.com

Noise Control
Edited by Marco Caniato and Federica Bettarello
p. cm.
Print ISBN 978-1-80356-011-3
Online ISBN 978-1-80356-012-0
eBook (PDF) ISBN 978-1-80356-013-7

Meet the editors

Marco Caniato is an Italian researcher and teacher. He is the author of more than 100 publications, including conference and journal papers and books. His main research is focused on the thermal and acoustic properties of construction and building materials as well as building elements. Another focus of his studies is the perception, evaluation, and implementation of indoor comfort for individuals with special needs. He is internationally involved in organizing conferences, managing projects, and coordinating scientific works.

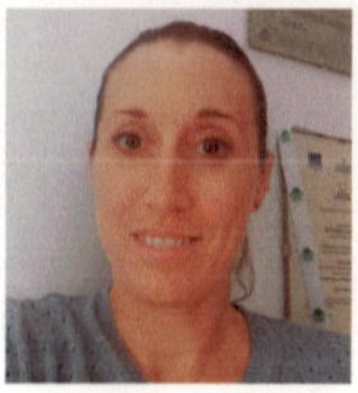

Federica Bettarello is an Italian researcher with a Ph.D. in Engineering Sciences (acoustics). Her research interests include building and architectural acoustics. She is the author of more than fifty conference and journal papers. She is involved in many international projects in which she coordinates acoustic activities. She also participates in international congresses and workshops.

Contents

Preface

Several research works on noise and noise control are currently available for the scientific, technical, and general public. **Figure 1** shows a graph of published papers on noise control from 2001 to 2021. As shown, the number of publications increased from about 3000 to about 9000 in the last 20 years. Is it necessary to publish more? Our answer is yes. We believe there will always be a need for more publications on this topic because noise control is constantly evolving along with human activities, technologies, and environments. The more these things change, the more noise will be present and need to be controlled.

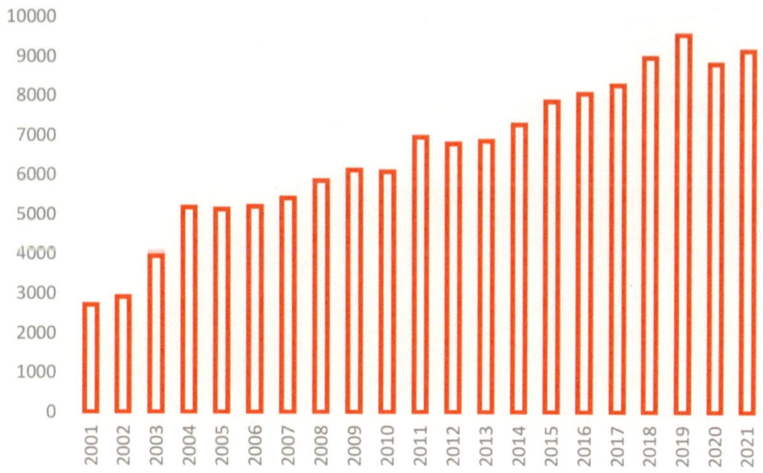

Figure 1.
Publication trends retrieved from Scopus for the years 2001 – 2021 using the keywords "noise control".

This book discusses different topics related to noise control, including noise control methods and techniques, noise pollution and solutions, and noise mapping.

Marco Caniato
Free University of Bolzano,
Bolzano, Italy

Federica Bettarello
University of Trieste,
Trieste, Italy

Chapter 1

Introductory Chapter: Is Noise Really Important? Should It Be Controlled?

Marco Caniato and Federica Bettarello

1. Introduction

Is noise really important? This is a very challenging scientific question. There are several literature works dealing with noise and its control. Anyway, why this physic phenomenon is so important for humans and specifically for scientists?

In years, researchers all over the word explored the entire acoustics field. If one writes "noise" on Google Scholar, almost 6 million results appear. This means that this parameter, this presence in our lives is of paramount interest and clearly occupies our minds and our thoughts.

Everybody knows that sound is a wave propagation in a dense mean. However, this is the academic definition. Aside of it, the perception is what makes the difference. And clearly what changes between *sound* and *noise*. The physics is the same, but sound is something positive such as music one likes or the voice of a beloved person. Noise is something disturbing such as industrial noise and traffic noise.

For these reasons, perception is what drives the research, rather than the physic phenomenon per se. Human perception varies from individual to individual. It is rather different, but some common trends are possible to find in literature. Thus, in the noise history, standards, scientific proposals, and law requirements were published, discussed, and imposed with the sole aim to control noise emission.

From what above reported, it is clear that the control of noise is needed for human purposes, for comfort reasons, and finally (and more importantly), for health issues and safety.

2. Noise control in literature

Literature is full of excellent works related to noise control. When one looks more in depth, some useful pieces of information can be derived. In order to do so, it is useful to picture the connection between the several different topics related to the "noise control" one.

When referring to **Figure 1**, it is possible to see that seven principal areas are related to the keyword "noise control," namely (in order of numerical importance given by literature papers:

1. noise related to electronic devices

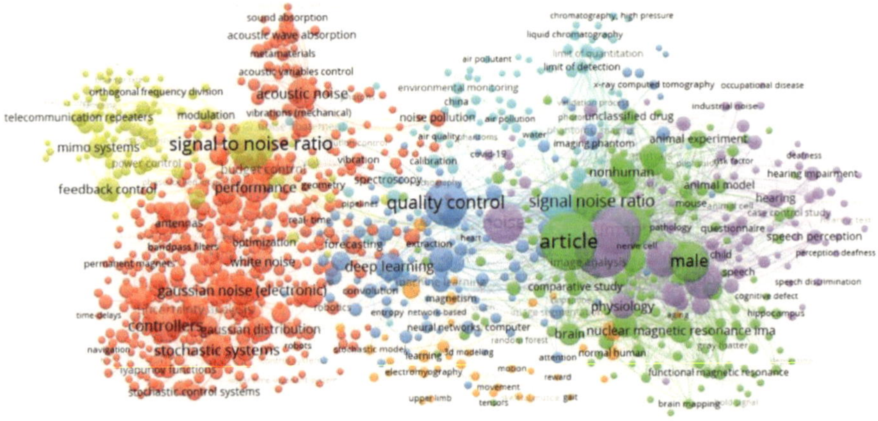

Figure 1.
VoS Viewer picture of the first 2000 records available in Scopus, using the keyword "noise control".

2. noise relate to outdoor environment

3. noise related to networks and connections

4. noise related to telecommunications

5. noise related to perception and impairment

6. noise related to simulations and numerical models

However, it is clear how many treated topics are not cross-linked together. As example, materials and metamaterials are only linked with vibrations and mechanics and not to the other fields. In the same way, speech perception and hearing impairment are related only one to each other but not for example with quality control, control systems, etc.

From **Figure 2**, it is interesting to notice how, when using a density analysis, the most important parameter related to noise control is related to human (yellow spot at the right), then to quality control (center), and then to controllers and how to control noise (left). When specifically focusing on noise as keywords, this cross-linked relation between "noise" and "human" topic becomes the most relevant (**Figure 3**). Here, the connection between noise, human, quality, and the psychological effect is assessed, confirming the initial hypothesis.

This demonstrates how more science is needed, especially cross-disciplinary research studies that make it possible to solve problems, to formulate new hypotheses, and to confirm or negate the ones that are still open.

The vision of this book is then to continue on the path of innovation in science considering different issues, apparently diverse one from another, but with a common idea: implement the knowledge on noise control and spread it.

With this in mind, different topics are here handled:

a. control techniques and methods [1]

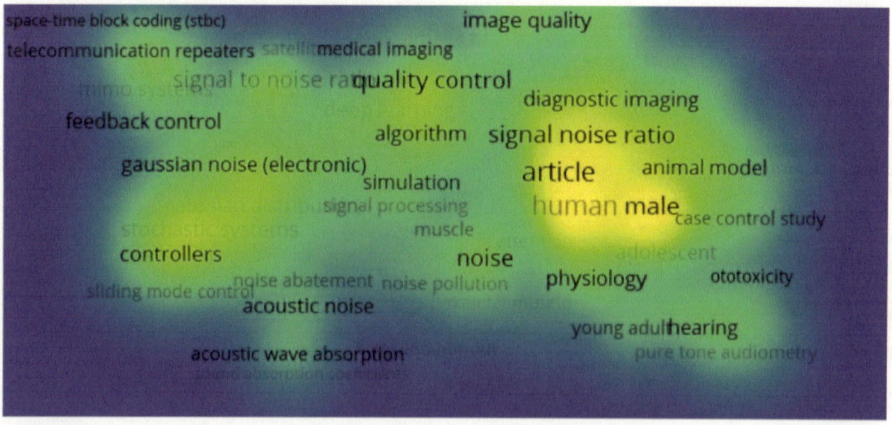

Figure 2.
VoS Viewer picture of the first 2000 records available in Scopus, using the keyword "noise control" – density visualization.

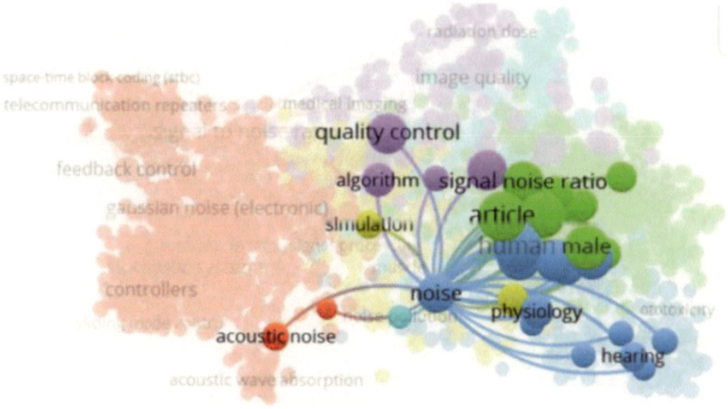

Figure 3.
VoS Viewer picture of the first 2000 records available in Scopus, using the keyword "noise control" – specific connections.

b. EVAC [2]

c. Pollution during COVID pandemic [3]

d. Outdoor Barriers [4]

e. Mapping [5]

3. Conclusions

It is here briefly demonstrated how noise is part of everybody's life and that its control is manly necessary because of human perception. Many fields of noise control are present in literature, but they are not well interconnected. This book proposes further food for thought in relation to this paramount issue, which is impacting our everyday life.

Author details

Marco Caniato[1*] and Federica Bettarello[2]

1 Faculty of Science and Technology, Free University of Bolzano, Bolzano, Italy

2 Engineering and Architecture Department, University of Trieste, Trieste, Italy

*Address all correspondence to: mcaniato@unibz.it

References

[1] González AE. Overview of noise control techniques and methods. In: Noise Control. London, UK: IntechOpen; 2022. Available from: https://www.intechopen.com/chapters/undefined/state.item.id

[2] Miyoshi T. Evacuation guidance assistance system using emitting sound. In: Noise Control. London, UK: IntechOpen; 2022. Available from: https://www.intechopen.com/chapters/undefined/state.item.id

[3] Swain BK, Das CP, Goswami S. Impact of Noise Pollution During Covid-19: A Case Study of Balasore, Odisha. IntechOpen; 2022

[4] Barba A, Martinez-Orozco JM. Approaches for Noise Barrier Effectiveness Evaluation Based on in situ "Insertion Loss" Determination. IntechOpen; 2022

[5] Ballesteros JA, Ballesteros MJ, Quintana S, Fernandez MD. Noise Profile Categorization for Noise Mapping in Cities: The Case of Cuenca (Spain). IntechOpen; 2022

Chapter 2

Overview of Noise Control Techniques and Methods

Alice Elizabeth González

Abstract

Noise control refers to a set of methods, techniques, and technologies that allows obtaining acceptable noise levels in a given place, according to economic and operational considerations. The question of "acceptance" is for what or for whom. Generally, there is no single answer to this question, nor is there a single solution to any given problem, as long as regulatory compliance is achieved. Noise control does not necessarily imply the reduction of noise emissions—it refers to making acceptable sound pressure levels of immission (i.e., the signal reaching the receiver). This chapter aims to present the basis of noise control techniques, both in emission and propagation, to finally achieve the most current protection techniques for the receivers, when there are no more alternatives in the previous steps.

Keywords: noise control, silencers, acoustic barriers, green barriers, active noise control

1. Introduction

When faced with a situation in which the sound pressure levels (SPL) are too high according to the intended use of a certain place, there are several options for enhancing its acoustic quality. Noise control solutions can be passive or active. Passive control is the most used solution; it aims to dissipate the excess acoustic energy through absorption, transmission, or diffusion. On the other hand, active noise control (ANC) solutions imply introducing a specific acoustic signal to counteract the sound waves to be controlled. As in most environmental issues, the possible measures must be analyzed seeking first to improve the performance of the source, that is, to reduce its emissions; then the possibilities of acting on the propagation path are to be studied; taking protection measures at the potentially affected receivers will be considered only as a last option for noise control.

Even with the latest technology to meet a requirement or to satisfy an objective, noise emissions can occur due to friction, impact, turbulence, imbalance of moving parts, and cavitation in fluids, among many other causes. In addition, the lack of maintenance or operating in an incorrect manner can increase noise emissions. Measures in the propagation pathway become necessary in many cases, for example, when working with high-power or high-speed machinery. The most common options are silencers, either reactive or dissipative; plenum chambers; acoustic coatings; enclosures; noise barriers; vibration dumping or insulation. When reducing the sound pressure levels at the receiver is still needed, the use of personal hearing protection is considered, although sometimes control booths or cabins can be very good options.

First, having an accurate diagnosis of the problem is essential. Some issues that should be considered before making a decision are ventilation requirements of the equipment; characteristics of location; availability of space; energy requirements; operation and maintenance; and costs.

This chapter presents the main features of passive control systems, which are usually the most common; and then, ANC solutions, which have various applications, especially for low-frequency noise.

2. Passive noise control

Passive noise control includes a wide set of techniques aimed to make the immission sound pressure level admissible to the receivers.

There are some general measures that can make a difference. They are low cost and based on common sense. It is rather common to find these improvement opportunities, for example, rearranging the work process to avoid unnecessary exposure to high noise levels to the public, changing the direction of loudspeakers at music venues, or encouraging the public to disperse at the end of recreation activities.

There is a wide spectrum of control measures to be considered, depending on the type of source. The sooner control measures are taken, the faster the results.

2.1 Engines noise control

Periodic preventive maintenance of devices, engine mobile parts, and machinery is essential—checking the condition of gears, bearings, and proper lubrication.

Noise emitted from internal combustion engines is the result of airflow processes into the engine (aerodynamic noise), from the mechanical movement of the engine (mechanical noise), and from pressure increases associated with the combustion process (combustion noise). Combustion noise is normally the predominant source in diesel engines, due to the rapid rise of pressure inside the cylinders. The noise emitted by a combustion engine is related to the efficiency of conversion of chemical energy into mechanical energy; the acoustic energy related to the roar of the engine is usually between 10^{-8} and 10^{-5} of its power. For two devices of the same power, the noise emitted by a diesel generator is greater than the one emitted by a natural gas one [1].

The acoustic power of engines is related to their maintenance—acoustic emissions are an indicator of waste of energy. For continuous combustion systems, the combustion noise can be expressed in terms of the thermo-acoustic efficiency, which is the ratio between the total energy of a sound pulse and the heat release rate. An engine in good operating conditions should normally emit between 10^{-6} and 10^{-5} of its power through acoustic energy; if the system is in improper condition, the acoustic emissions could rise to 10^{-4} of its thermal power [2]. The maximum thermo-acoustic efficiency expected for unconfined hydrocarbon flames is 10^{-6}. When a combustion engine is not in proper operating conditions, the frequencies that denote higher sound pressure levels are the harmonics of the rotational frequency. This also happens in other rotating or reciprocating machines. Malfunctions in electric motors are usually related to excessive noise in harmonics of the synchronous frequency. In other electrical devices, the noise appears in harmonics of the line frequency. A catalog of engine problems and the frequencies where they appear is presented in Ref. [3].

Heat recovery boilers are recommended as noise control devices for large engines. They act as passive silencers, but when installed, of course, they can provide other services too (e.g., heating for decreasing the fuel viscosity).

A heat recovery steam generator performs "*a secondary function as an in-line silencer for combustion turbine noise emissions*" [4]. The recovery boiler can be considered part of the noise emission control system to comply with the immission level regulations. Hence, it must be specified as such.

Sometimes dedicated passive silencers can be avoided by the installation of recovery boilers with a secondary function as exhaust silencers. **Table 1** presents the reduction in SPL measured between up-flow and down-flow of a recovery boiler. The reduction is from 10 dB (at 125 Hz) to 35 dB (at 8000 Hz). The reduction in A-weighted SPL is also high: 25 dB [5].

2.2 Passive silencers

Passive systems, whose generic designation is silencers, can act either through their geometric characteristics or through the incorporation of acoustic absorbent materials. According to their principle of action, silencers can be classified into two families: reactive or reflective silencers and dissipative or resistive silencers.

2.2.1 Reactive or reflective silencers

The principle operation of reactive silencers is based on generating sound reflections from geometric properties of the propagation medium, for example, section changes. They are usually solved with coupled tubes. **Figure 1** shows a sketch of a reactive silencer; how to calculate its transmission loss (TL) is also presented.

For a discontinuous area reactive silencer as sketched:

When an acoustic wave of amplitude A_i propagates with velocity v_1 in a tube of section S_1 and it changes abruptly to a section S_2, a reflected wave of amplitude A_r goes back to the source and a transmitted wave of amplitude A_t continues propagating with velocity v_2. According to Snell's Law, the amplitudes of these waves (A_r and A_t) can be written as [6]:

$$A_t = \frac{2\,v_2}{v_1 + v_2} A_i; A_r = \frac{v_2 - v_1}{v_1 + v_2} A_i \tag{1}$$

f (Hz)	63	125	250	500	1000	2000	4000	8000	A-weighted
Reduction (dB)	11	10	18	25	28	26	31	35	25

Table 1.
SPL reduction achieved by a recovery boiler in a paper mill (from Ref. [5]).

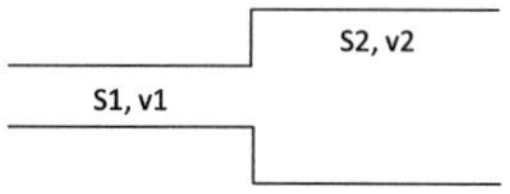

For a discontinuous area reactive silencer as sketched:

$$TL = 10\log\left[\frac{(S_1 + S_2)^2}{4\,S_1 S_2}\right]$$

Figure 1.
Sketch of operation principles of reactive silencers (adapted from Ref. [7]).

The greater the difference between S_1 and S_2, the better attenuation is obtained; the shorter the transition between S_1 and S_2, the better attenuation is also obtained. According to Ref. [6], the SPL reduction in an abrupt expansion as sketched in **Figure 1** could be between 2 dB (for $S_2/S_1 = 4$) to more than 8 dB (for $S_2/S_1 = 25$).

These silencers do not require absorbent materials. Their best application is to control narrow band noise or tones at discrete frequencies, especially low frequencies. The most common designs are sketched in **Table 2**. The highest frequency of plane (or bidimensional) waves to be controlled by this type of silencer can be estimated as [6]: $f < c/2a$, where c is the velocity of sound and a is the diameter of the tube, or the minor dimension of its section if it is rectangular. It is important to consider that the wavelength λ of the waves to be controlled should also satisfy: $\lambda \ll a$.

Also, the well-known Helmholtz resonator can be designed to serve as a silencer, for example, mounting a Helmholtz device designed according to the frequencies to control one side of the tube/duct, where the noise control is needed. A reduction up to 8 dB can be obtained [7]. The volume V of the cavity and the radio r and length l of the neck are to be determined according to the frequency f to control, using the equation of the resonance frequency of a Helmholtz presented in **Figure 2**.

2.2.2 Resistive or dissipative silencers

These are the most suitable silencers to control medium and high frequencies sectors of the broadband noise spectrum. These consist of devices coated with a porous acoustic absorbing material, in contact with the airflow where noise control is needed. The most usual acoustic absorbing materials for the coats are mineral wool or fiberglass. Their effectiveness is based on dissipating the acoustic energy of sound waves as heat, by forcing the flow to lose its energy through friction.

The performance of acoustic dissipative silencers depends on the features of the coating, the thickness and length of the acoustic absorbing materials, the features of

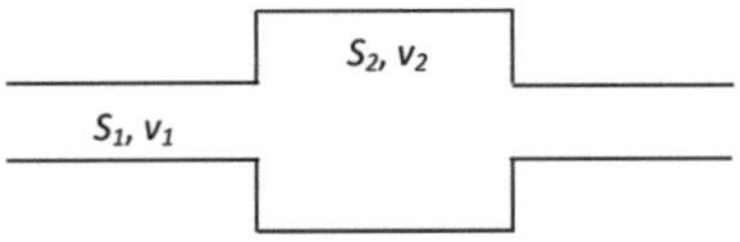	Expansion chamber: $TL = 10 \log \left[1 + \left(\frac{S_1^2 - S_2^2}{2S_1S_2}\right)^2 \cdot sen^2\left(\frac{2\pi l}{\lambda}\right)\right]$
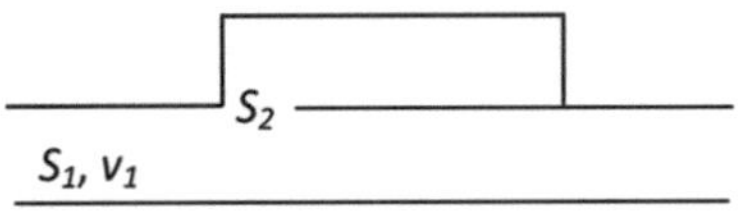	Side branch reactive silencer: $TL = 10 \log \left[\frac{S_1 \cdot S_2 \cdot (S_1 + S_2)}{4}\right]$

Table 2.
Sketch and TL of two designs of reactive silencers (adapted from Ref. [7]).

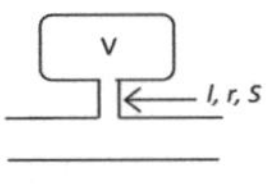	$f = \frac{c}{2\pi}\sqrt{\frac{S}{V\,l_e}}$ S is the area of the cross section of the neck l_e is the effective length of the neck: $l_e = l + 1{,}7\,r$

Figure 2.
Sketch of a side Helmholtz silencer (adapted from Ref. [6]).

the flow passage section, and the pressure drop in the system (remember that the pressure drop is proportional to the squared velocity of the flow u^2).

2.3 Encapsulations

Encapsulations can be suitable options for noise control for many types of equipment. The most used enclosures are small buildings where the noisy machines are enclosed. They must be designed by taking into account not only the acoustic characteristics but the ventilation needed for proper operation. Even when each machine complies with noise emission standards, the need of reducing SPL in areas close to public or private works is a major concern. Giraldo Arango states that "Empresas Públicas de Medellín E.S.P." (Public Companies of Medellín, Colombia) successfully incorporated some environmental management requirements in its work specification, especially for Public Works [8]:

> *… the use of modern machinery and equipment with soundproofed engines; preventive and corrective maintenance measures must be taken to keep equipment in proper working conditions; if not possible to insulate the sound emission of, that is, jackhammers or disc cutters, then a maximum of two continued hours of noisy equipment operation should be taken, with breaks of the same duration; coordinating and scheduling with authorities of schools and health institutions, in order to conduct noisy operations during class breaks or shift changes …*

He also states: "*… people from public and private sectors mention it is not possible to control noise. … it has been proven that it is possible in practice by using other methods as those normally used, and which are environmentally and economically more attractive.*" In his opinion, stricter control of environmental management is needed; control activities should be handled by the contractor [8].

Some machines have the possibility of building a customized enclosure, either removable or fixed. They must have both good acoustic insulation and absorption properties. **Table 3** presents the comparison of the SPL measured at 1 m from two twin well-point pumps [9]. One of them was enclosed in a customized acoustic removable capsule. Both measurements were done with a class 1 sound level meter during the same morning when the pumps were working in similar conditions in the same area of the city. Drastic reduction in the values of some parameters was experienced for A-weighted sound pressure levels.

2.4 Traffic noise control: special pavements

The noise of road vehicles is generated from three main sources—engine acoustic emissions, which represent the main source at low velocity; tire-pavement noise, which is dominant from 60 to 100 km/h (approximately); and aerodynamic noise, which is more significant at high speeds (greater than 100 km/h).

Acoustic pavements are designed for reducing noise due to contact with the tires. They have a high sound absorption coefficient in the frequency range where rolling emissions are greater. Thus, the acoustic energy is dissipated as heat, rather than reflected on the ground. Smooth pavements will be less noisy at high speeds if they are dry; in wet conditions, an increase of about 4 dB can occur.

There are different responses of pavements according to the frequency. The results depend on the mass percentage of the asphalt, the air pockets in mineral aggregates, and

	Non-enclosed machine	Enclosed machine	Difference	Observations
L_{AFmax}	87.2	77.1	−10.1	Significant attenuation
L_{AFmin}	82.6	74.3	−8.3	Significant attenuation
L_{AIeq}	86.4	76.4	−10.0	Significant attenuation
L_{AFeq}	84.8	75.5	−9.3	Significant attenuation
L_{CFeq}	91.1	90.2	−0.9	As expected, the enclosure is not effective at low frequencies
L_{AF10}	85.4	75.9	−9.5	Significant attenuation
L_{AF50}	84.8	75.5	−9.3	Significant attenuation
L_{AF90}	84.1	75.1	−9.0	Significant attenuation
$L_{AIeq} - L_{AFeq}$	1.6	1.0	−0.6	As L_{AFeq} and L_{AIeq} present similar decreases, significant changes none were found
$L_{CFeq} - L_{AFeq}$	6.3	14.8	+8.5	Due to the ineffectiveness of the enclosure at low frequencies, L_{CFeq} does not change significantly, while L_{AFeq} does. Hence, the difference $L_{CFeq} - L_{AFeq}$ raises
$L_{AF10} - L_{AF90}$	1.3	0.8	−0.5	As L_{AF10} and L_{AF90} present similar decreases, significant changes none were found

L_{AFmax} = *A-weighted maximum sound pressure level, measured in fast time weighting;* L_{AFmin} = *A-weighted minimum sound pressure level, measured in fast time weighting;* L_{AIeq} = *A-weighted equivalent sound pressure level, measured in impulse time weighting;* L_{AFeq} = *A-weighted equivalent sound pressure level, measured in fast time weighting;* L_{CFeq} = *C-weighted equivalent sound pressure level, measured in fast time weighting;* L_{AF10} = *A-weighted 10% exceedance sound pressure level, measured in fast time weighting;* L_{AF50} = *A-weighted 50% exceedance sound pressure level, measured in fast time weighting;* L_{AF90} = *A-weighted 90% exceedance sound pressure level, measured in fast time weighting.*

Table 3.
Comparison of the acoustic performances of two well-point pumps, one of them into a customized acoustic enclosure (from Ref. [9]).

the aggregates grading curve [10]. The most silent pavements are the so-called draining or porous pavements. They can reduce up to 5 dB, especially in high frequencies. They have a very high percentage of holes in their structure, which absorb part of the sound energy emitted by vehicles (both by the tires and by the engine) and drain rainwater as well. The high porosity is obtained by using uniform-sized aggregates.

Acoustic pavements are not only expensive but also they age as their pores clog; then, they lose their acoustic properties. Highways and high-speed roads are less vulnerable to aging, due to a self-cleaning effect caused by the movement of vehicles at high speeds. In urban areas where circulation speeds are slower, permeability is quickly lost. The cleaning procedures are expensive, difficult and of limited efficiency, so the construction of these special pavements is restricted to high-speed traffic lanes.

There are currently promising developments in acoustic pavements that use reclaimed materials. The aging phenomenon was studied at three pavement sections of rubberized asphalt (i.e., asphalt containing crumb rubber from tires) [11]. The best analytic relation to link acoustic properties and aging is not linear but logarithmic. The main variables that correlate in a direct sense are the temperature of air and pavement, and the hardness of the rubber of the tires; an inverse relation was found with other variables, such as heavy traffic flow, age of the pavement, and climate variables [11].

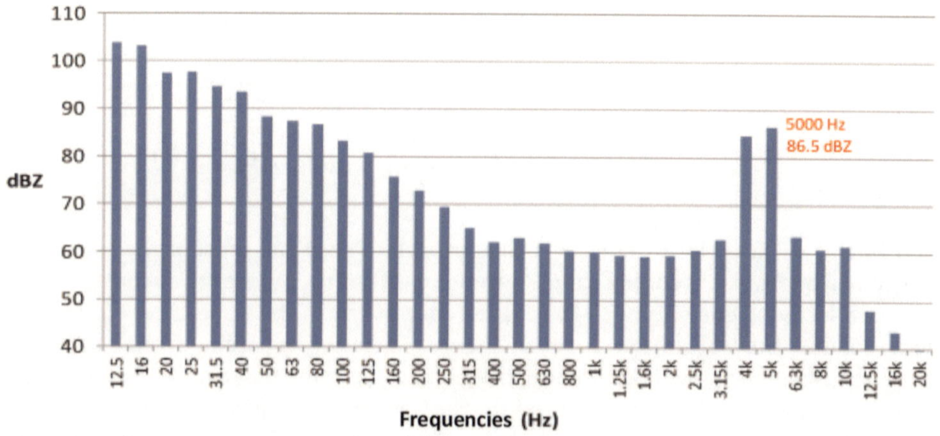

Figure 3.
Measured sound pressure level spectrum of a wind-induced annoying noise caused by airflow through small holes (from Ref. [12]).

2.5 Aerodynamic noise

The interaction between wind and constructions can produce acoustic emissions. Sometimes they are caused by the detachment of the boundary layer developed on a surface, for example, the blades of a wind turbine, the hood of a vehicle, or the wings of an aircraft. The use of a slitted-sawtooth serrated trailing edge to control the trailing edge noise in large wind turbines reduces SPL by about 5 dB [13]. When the noise is related to the airflow through holes or slots, the noise spectrum usually presents strong pure tones, as shown in **Figure 3**.

The emission of noise can be related to a constriction in the flow that causes an increase in the velocity of the airflow; it is the case of the passage of air through holes, slots, or openings that are part of the design of a building. Analyzing the geometry of the problem and the statistics of wind velocity and direction, the most suitable solutions for preventing the phenomenon are—modifying the dimensions of holes and slots; selectively blocking or covering the openings [14]; or modifying the geometry of the wind passage to act on the degree of turbulence of the incoming flow [12].

3. Passive noise control along the propagation path

The intensity of a sound wave is the acoustic energy carried by the wave per unit of area and per unit of time. It can be computed as the relation between the squared sound pressure and the acoustic impedance of the propagation medium. Thus, two possible ways to reduce the sound intensity are acting directly on the sound pressure or acting on the acoustic impedance along the propagation path.

3.1 Changes in acoustic impedance along noise path

A change in the acoustic impedance Z along the sound path imposes a modification on the sound wave amplitude. When a sound wave intends to pass from a propagation medium **1** with acoustic impedance Z_1 to another medium **2** with acoustic impedance

Z_2, the fraction of its intensity transmitted from **1** to **2** is determined by the transmission factor F_t. The non-transmitted energy goes back to medium **1**, according to the reflection factor F_r. These factors can be written as relations between Z_1 and Z_2 [6]:

$$F_t = \frac{4\,Z_1Z_2}{(Z_1+Z_2)^2};F_r = \left(\frac{Z_1-Z_2}{Z_1+Z_2}\right)^2 \text{ with } F_t+F_r=1 \tag{2}$$

If Z_1 and Z_2 have close values, F_r will approach to zero and most of the acoustic energy will be transmitted. If Z_1 and Z_2 are very different, most of the acoustic energy will be reflected. Generalizing, a large difference between the impedances of the two media reduces the transmitted acoustic energy. Then, when passing from **1** to **2**, most of the acoustic energy will not be transmitted and the same will occur when passing back from **2** to **1**. This principle is used for the design of composite materials for acoustic insulation, but it is also suitable for controlling solid-induced noise and vibrations. In the first case, the best materials for acoustic insulation of airborne noise are those with very high acoustic impedance, because the acoustic impedance of the air has a low value (around 415 rayl).

When vibrations and/or solid-induced noise are to be controlled, one of the preferred solutions is "cutting" the propagation path and filling the joint with a soft material whose acoustic impedance value is as far as possible from that of the transmission medium. It is a common solution when building double walls to have independent foundations, but also the principle of floating floors and a good option for damping/insulating floor- or wall-transmitted vibrations.

Table 4 presents the value of longitudinal sound velocity, density, and acoustic impedance for some materials, including some types of common polymers.

3.2 Acoustic barriers or screens

The main parameter to describe the performance of acoustic barriers is the insertion loss (IL), which represents the difference between the SPL at a receiver without and with the barrier. The maximum value of IL that can be theoretically achieved is 20 dBA for thin screens and 23 dBA for earth embankments [15].

3.2.1 How does an acoustic barrier work?

An acoustic barrier or screen consists of an obstacle—usually, similar to a wall—that stands between a sound source and a receiver, and whose characteristics are defined to acoustically protect the receiver. The length of the screen normal to the source-receiver line may be greater than the wavelength for which the barrier is designed. Acoustic barriers aim to create a relatively calm and silent space behind it, in the so-called "acoustic shadow area," even at a short distance from any relevant sound source. The most frequent applications of acoustic barriers are concentrated around highways and railways, in construction sites and mining areas, in the vicinities of airports and industrial zones. They may be placed as close to the source as possible, to maximize their performance. Even though, sometimes it is necessary to put them close to the neighborhoods to protect them against the noise from different sources.

Four types of phenomena occur in an acoustic barrier: reflection, absorption, refraction, and diffraction (**Figure 4**).

Medium	Sound velocity c [m/s]	Density ρ [kg/m^3]	Acoustic impedance Z [rayl]
Aluminum	6300	2700	16.8×10^6
Bone	4080	1900	7.7×10^6
Brick	4300	1700	7.3×10^6
Cellulose butyrate	2140	1190	2.55×10^6
Clay	3400	2600	8.84×10^6
Concrete	3500	2600	8.9×10^6
Dry sand	1700	1610	2.74×10^6
Ethyl vinyl acetate (EVA)	1800	94	0.17×10^6
HDPE plate	2570	950	2.4×10^6
Iron	5900	7690	46×10^6
Iron cast	4600	7220	33.2×10^6
LDPE plate black	2180	915	1.9×10^6
Lead	1190	11,300	13.5×10^6
Marble	3810	2800	10.5×10^6
Neoprene	1600	1310	2.1×10^6
Nylon (polyamide)	2710	1150	3.1×10^6
Paraffin	1940	910	1.76×10^6
Polymethylmethacrylate (PMMA)	2724	1180	3.2×10^6
Polycarbonate	2250	1210	2.77×10^6
Polyester casting resin	2290	1070	2.5×10^6
Polyoxymethylene (POM)	2430	1420	3.45×10^6
Polypropylene	2470	880	2.2×10^6
Polystyrene rexolite	2346	1040	2.3×10^6
Polyurethane elastomers	1700	1040	1.80×10^6
Polytetrafluoroethylene (PTFE)	1530	2200	3.4×10^6
(PVC)	2380	1380	2.9×10^6
Recycled plastic	857	940	0.8×10^6
Rubber-like material	1300–1700	900–1300	$(1.2–2.2) \times 10^6$
Stainless steel	5800	7930	46×10^6
Tin	3300	7300	24.2×10^6
Vinyl rigid	2230	1330	2.96×10^6
Wood	4000	700	2.8×10^6
Wood cork	500	250	0.12×10^6
Zinc	4200	7000	29.4×10^6

Table 4.
Acoustic properties of some materials (values from various sources).

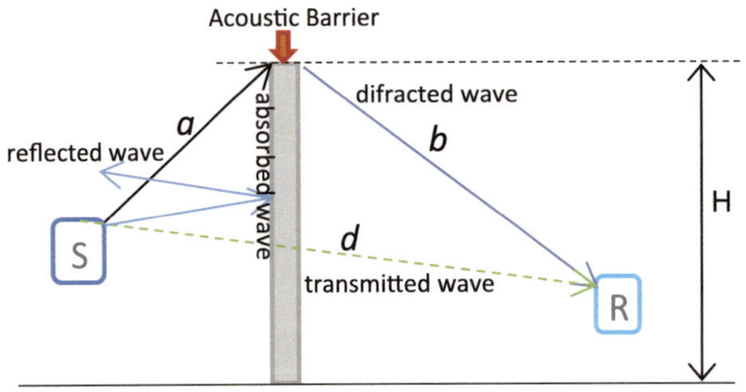

Figure 4.
Acoustic phenomena that occur in a barrier (adapted from Ref. [16]).

Reflection and refraction follow Snell's law; the fraction of the acoustic energy that will be refracted and reflected can be estimated with the coefficients F_r and F_t presented in the previous section. Please consider that the refracted energy involves both the transmitted and the absorbed at the surface of the screen, that is, the refracted energy is all the energy that is not reflected.

The requirements for the materials for building acoustic barriers are not acoustically challenging: a minimum surface density of 10 kg/m^2 is sufficient to obtain an adequate transmission loss (TL) value. Tightness is important: a percentage of openings of more than 1.5% in the screen surface causes a reduction of 3 dB in its transmission loss TL.

The most important feature to consider for designing a noise barrier is diffraction. Both the upper and the side edges of a noise barrier become diffracted noise sources. Hence, for calculating the SPL at a receiver without the direct vision of the main noise source, the contributions of each one of these new sources are to be added. If the barrier has a special heading, its reduction may be also considered.

A barrier can be considered thin or thick. A thin barrier can be thought of as a flat surface with mass but without thickness; thus, its edges would act as lines where diffraction can occur. An acoustic barrier is said to be thick when it has more than one point where diffraction can occur. The most frequent case is that of embankments, although the buildings of a city are very important to control noise pollution, as they have many edges for diffraction.

When the wavelength to be controlled is less than 20% of the top width ($\lambda < e/5$), the barrier will be considered thick. If the top width exceeds 3 m, the barrier will behave as thick for all frequencies. Otherwise, the barrier will behave as it is thin and it should be designed in such a way. In the case of thick screens, the thickness e may be added to the smallest distance a or b (**Figure 5**) to get the new values a' or b'. All calculations may be computed using the new values a' or b'.

A thin barrier can turn into a thick one simply by adding a proper header for having multi-edge diffraction and enhancing its performance and effectiveness. Using absorbing materials on the surface exposed to the noise source is not mandatory; nevertheless, it can be useful for reducing noise reflections, that is, when there are screens on both sides of a highway or railway.

The best geometry for using absorbing materials is when the height H of the screen is greater than 20% of its length L ($H > L/5$). In addition, the best solution for avoiding reflections when $L/10 > H > L/20$, is by positioning one of the two screens at a minimum angle of 15° vertical. When $L/5 > H > L/10$, cases should be studied individually. If $H < L/20$, no simple action will significantly improve its performance [18].

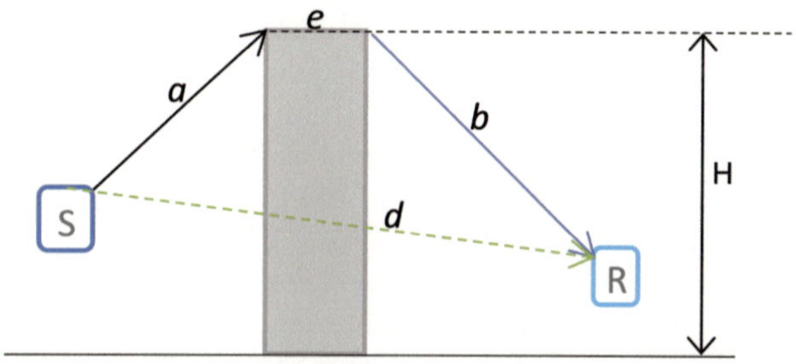

Figure 5.
Cross section of a thick screen (adapted from Ref. [17]).

Many natural fibers can contribute to acoustic absorption; the best performance was achieved by fibers, such as kapok, pineapple-leaf, and hemp [19].

Substituting a reflective barrier placed 1 m far from a railway with an absorptive one could improve IL up to 6–10 dB [20]. Placing a Schroeder diffuser on the surface of an extruded-PVC acoustic barrier could improve the IL in almost all frequencies between 315 Hz and 8000 Hz [21].

When a transparent or a reflective material is selected for a noise barrier, the birds will not be able to visually recognize it. Thus, to minimize bird collisions onto barriers, the use of opaque stripes or dots should be taken into account in the terms of reference [22]. The old usage of painting raptors onto transparent barriers [20] can also be substituted with an ultraviolet "A" reflecting coating (UV-A), that is visible only to birds.

3.2.2 Types of acoustic barriers

When selecting the type of acoustic barrier to construct, not only the acoustic performance must be taken into account, but also the available space, characteristics of topsoil and subsoil conditions, the possibilities of landscaping integration, and the cost per m or per km, both for construction/mounting and maintenance.

The most common types of barriers are thin screens, soil embankments, partitions or enclosures, and green curtains:

- Thin screens: They are relatively thin constructions for controlling noise in the vicinity of roads or railways. Their IL strongly depends on their height, length, and materials. Thin screens can be built of concrete, absorbent bricks, galvanized steel, wood, wood panels, aluminum, methacrylate, and polycarbonate, among others.

- Soil embankments: They are mounds that are usually covered with grass. They usually have a good acceptance. Their main disadvantages are the need for huge space for their construction and constant maintenance. Nevertheless, when there is enough space and proper materials for its construction, it is a rather economical solution. The IL should be calculated as a multiple-diffraction screen (thick barrier). It should be noted that the height of a ground embankment used as an acoustic barrier should be higher than a thin barrier to achieve the same results.

- Partitions or enclosures: They are partial or total covers of a route or street section, to avoid noise propagation. They are very effective, but their construction and maintenance can be very costly.

- Green barriers: They produce the best esthetic result. On the other hand, their acoustic performance is poor and they need constant maintenance.

Both thin and green barriers can be designed as modular infrastructure, to build different configurations [23]. The "A-frames" built out of Corten steel facilitate plants growth, protection, irrigation, and maintenance [20].

The costs of different materials of noise barriers by life-cycle cost analysis are compared in Ref. [24]. The present net worth is considered by adding the initial construction cost and the annual cost of maintenance and replacement during a life of n years, with an interest rate i of 5.5%. The service life of embankments and concrete is considered of n = 50 years, and of n = 25 years for steel-, wood- and aluminum-based materials. In this framework, the most cost-efficient option was earth embankments; the least cost-efficient was the aluminum-based materials (four times more expensive than earth embankments).

3.2.3 Basic acoustic design of an infinite acoustic barrier

Although there are several explicit calculation methods, they are only approximations. If an accurate value is needed, applying numerical modeling techniques is mandatory. Possibly, one of the earliest methods to calculate the IL of an acoustic barrier is Maekawa's [20]. It considers frequencies between 100 Hz and 5000 Hz. The depletion of SPL due to diffraction on the top edge can be calculated as IL = 10 log(20 N).

N is the Fresnel number and it can be obtained with basis on the wavelength λ and the difference between the direct path d (the geometric distance between source and receiver) and the path over the barrier, or diffracted path (a + b), as indicated in **Figure 4**:

$$N = \frac{2}{\lambda}(a + b - d) = \frac{2\delta}{\lambda} \tag{3}$$

Best results of Maekawa's expression are obtained when the height of the barrier is significantly less than its length, a is less than 5 m, the height of the source is greater than a and the height of the receiver is greater than b.

Other expressions for estimating IL are in use, aiming to improve the accuracy of the results [20, 25]:

- For point sources: IL = 10 log(3 + 20 N) with $N > 1$.

- For linear sources and traffic noise: IL = 10 log(2 + 5.5 N) with $N > 1$.

Kurze-Anderson [20]:

$$\Delta L = \mathrm{IL} = 20\ \log\frac{\sqrt{2\pi N}}{\tanh\sqrt{2\pi N}} + 5\ \text{for} - 0.2 < N < 12.5 \tag{4}$$

or ΔL = IL = 24 dB for $N > 12.5$ (only for point source).

Taking into account that $tanh\, X = \frac{senh\ X}{\cosh X} = \frac{e^X - e^{-X}}{e^X + e^{-X}}$.

When ground absorption is to be considered, please use the next expression, either for thin or thick barriers:

$$\mathrm{IL} = 10\ \log\left(3 + 10\ N\ K\right) - A_{ground} \tag{5}$$

The ground absorption A_{ground} can be computed by any proper method.
The value of K is related to meteorological conditions, according to:

When $100\ m < a + b < 300\ m : K = e^{-0.0005\sqrt{\frac{abd}{N\lambda}}}$; otherwise, $K = 1$

Pita Olalla [26] proposes the following expression for a linear source:

$$IL = \Delta L = 15\log\frac{\sqrt{2\pi N}}{tanh\ \sqrt{2\pi N}} + 5 - 10\ log\left(2e^{-h/2\lambda} + 1\right) \tag{6}$$

where h = height of the acoustic barrier.

3.2.4 Minimum length of a noise barrier

To calculate the minimum length of a barrier, the most important variable is the angle of view of the receiver (**Figure 6**). If the angles of view of the receiver are α and β, the acoustic energy that can reach it will be proportional to the ratio between ($\alpha + \beta$) and the straight angle of which they are part. Thus, if the receiver is located at a distance d from the road, the expected sound pressure level at its location will be [18]:

$$L_{p,d} = L_d - IL + 10\ \log\frac{180° - (\alpha + \beta)}{180°} \tag{7}$$

where $L_{p,d}$ is the SPL at distance d from the source, with the screen; L_d is the SPL at the same point, without the screen; IL insertion loss of the noise barrier, if infinite; and α, β are the angles of direct view of the source from the receiver, at both sides of the screen.

The relation between the IL and the shielding angle from the receiver does not depend on the distance from the receiver to the barrier, as stated by Ross et al. [27] after analyzing a set of 300 sites with barriers from 90 to 1100 m in length. When studying the behavior of barriers of the same length and different heights, the shielding angle should be at least 165° to reach a good IL value. The effectiveness of the barrier will have an abrupt depletion if the shielding angle is less than 140°. These results backup the usual recommendation of having a shielding angle of at least 160°, which is rather the same as having a barrier length of at least four times the distance d on each side of the receiver to be protected. (Strictly, the value should be closer to / between 6–7.5d).

3.2.5 Headers or cappings

While the basic design of a noise barrier involves defining the location, height, and materials of a wall with a horizontal top edge, there are many possible designs for capping the wall. Headers or cappings are very important in the acoustic performance of barriers. Although they also have an esthetic function, headers seek to improve

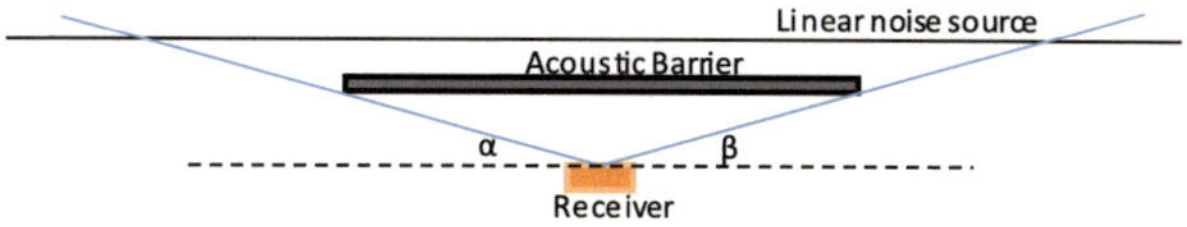

Figure 6.
Sketch for calculating the length of an acoustic screen (adapted from Ref. [18]).

noise reduction achieved by diffraction at the upper edge of the screen. There are cantilevered cappings, multi-diffraction, tubular, Y-shaped, absorptive or reflective ones, designs for promoting destructive interference, sound diffusion or scattering. There are even patented acoustic header designs for acoustic barriers. A broad catalog of tested header designs can be found in Ref. [26]. The excess attenuation is near 2 dB, but an extra attenuation of up to 6 dB can be achieved. When comparing the performance of two thin barriers of the same height in the same construction site, one of them having a straight edge cantilever and the other one, a slanted flat-tip jagged cantilever, the last one achieved an additional attenuation of up to 5 dBA [28]. Last but not least, there are many designs of cappings that includes ANC technologies.

3.3 Green barriers

A green barrier designed to have an acoustic function should not be confused with an esthetic one. In fact, guidelines for greening acoustic barriers can be found, for example [29].

A study carried out in Virginia (US) concluded that "*there was minimal noise attenuation that could be attributed to the coniferous trees at the 15 study sites examined. Attenuation was not correlated with tree stand age, height, species, or density for these sites.*" [30]

Shrubs and trees should meet certain characteristics to take part in a noise barrier. The depth of the green curtain may be at least 15–40 m. Even if its best performance is between 250 and 1000 Hz, an attenuation greater than 3 dB in L_{Aeq} should not be expected. The best attenuation performance of a green barrier occurs at a wavelength twice the size of the leaves of the tree species [20].

The expression for estimating the attenuation of a dense green barrier, proposed by Kurze, Beranek and Hoover, as presented in Ref. [31]:

$$\Delta L\ [dB] = \frac{d\ [m]}{100} f^{\frac{1}{3}} \tag{8}$$

Even if direct relations between the IL and the length, depth and diameter of the branches of urban green curtains have been reported and drawn [32], analytic expressions for predictive purposes are not easy to develop.

It is usually assumed that the attenuation of green barriers is almost negligible for frequencies below 400 Hz and that they may be useful only for frequencies above 1000 Hz, provided that the curtain is dense and its thickness is several tens of meters in depth. The improvement in the attenuation in low-frequencies is mostly related to fallen leaves and branches, which increases the surface porosity and thus, the ground absorption. Martens' work shows the best performance is reached when the leaf dimension is similar to half of the wavelength to absorb [20].

In a study analyzed in Ref. [31], the acoustic attenuation of two green barriers of cedar trees 4–4.5 m in height is compared. In the first case, the trees were aligned by rows and columns, in an area of 30 m × 5 m, very close to one another, with branches overlapping; in the other case, The trees were planted in a diagonal pattern, on the vertices of equilater triangles of 1.20 m side (quincunx pattern)., occupying an area of 25 × 9.20 m. The results are shown in **Table 5**.

The absorption coefficient of leaves and plants seems to be low. Low absorption coefficients were found up to 1600 Hz, most of them no greater than 0.30. The best performance was that of Winter *Primula vulgaris*, with values between 0.6 and 0.7 for frequencies of 500–1600 Hz; and the worst, *Hedera Helix's*, with all values below 0.20.

	Rows and columns	Diagonal
630 Hz	1.2	0.4
2000 Hz	2.6	5.2
5000 Hz	5.5	—
10,000 Hz	8.3	11.3

Table 5.
Attenuation of a green barrier of cedar trees in dB/m of curtain depthness in two different arrangements (from Ref. [31]).

The study states that "*the leaf area density and dominant angle of leaf orientation are two key morphological characteristics that can be used to predict accurately the effective flow resistivity and tortuosity of plants*" [33]. Two other studies obtained similar results [34, 35]. The main conclusion of Asdrubali et al. [34] was that "*the main absorber is the substrate soil (…). The presence of the plants becomes useful only when a large number of them is installed on the sample, otherwise is even pejorative within some frequency ranges.*" Just the opposite, Azkorra et al. [36] obtained a weighted sound absorption coefficient of 0.40, but the best absorption coefficients were at frequencies of 125 and 4000 Hz, and the worst ones, were at 500 and 1000 Hz.

3.4 Sonic crystal (acoustic open barriers)

Sonic crystals (SC) are periodic structures that have been studied in recent decades to learn about their possibilities of behaving as "acoustic open barriers". They have been also been studied for using them for sound diffusion [37].

The idea of an open barrier seems incompatible with the tightness and continuity of the surface of a "conventional" barrier. A SC is a periodic structure that can be triangular, squared, rectangular, or hexagonal; the best geometry is triangular [16]. The essential point is that the structure is the same regardless of the orientation of the material. Depending on the number of dimensions in which the pattern is repeated, the SC will be one-dimensional, two-dimensional, or three-dimensional. The most useful structure for acoustic barriers is two-dimensional with Cermet's topology (i.e., there is no contact between one scatterer to the other, each scatterer is totally surrounded by the external fluid—the air in this case) [38]. Thus, the key is that the operation principle is different in nature—the SC does not attenuate noise by diffraction, but by Bragg interference, widely used in optics—when a wave reaches a crystal structure surrounded by a fluid, the energy of some frequencies cannot be transmitted. These frequencies are the so-called "*Bragg frequencies.*" This occurs due to a destructive interference caused by a multiple scattering process [16, 37–39]. If a is the distance between two scatterers, the main controlled frequency will be $f = c/2a$ (c is the sound velocity and a is the so-called *constant of the lattice*). Due to the destructive interference caused by the SC structure, all the acoustic energy at that frequency is reflected. This frequency itself is a *bandgap*; the width of the bandgaps is linked to the *filling factor*, which represents the fraction of the lattice area that is occupied by scatterers. The greater the filling factor—without losing the crystal structure—the best IL is achieved; but if the filling factor is close to 1, the SC will become a continuous barrier [38]. It is worth mentioning that the height of the barrier is not a critical factor in the design [16].

The main advantages of acoustic open barriers are that they are lighter, easier to install and they do not need to have a significant additional height above the source. Research on this technology in the last years has allowed the development of new metamaterials, for example, a scatterer consisting of an absorbent split ring resonator, which has the structure of a SC with a cavity to act as a Helmholtz resonator and an absorbing material to reduce the acoustic energy at high frequencies [38].

4. Passive control at the receivers

It refers to actions aimed at improving acoustic insulation in recipients' homes, through constructive solutions, materials with better acoustic performance, windows sealers, special openings, or with double glass. When none of these options seems to be a solution, the use of personal protection equipment (PPE) is the last option. PPE are not only used by workers: many people who are very sensitive to noise have to wear these devices to sleep, or at least to fall asleep.

Even though, there is another way to improve sound quality for the recipients: it consists of masking the undesirable signal. Even if it is closer to an active method for noise control—it consists of adding acoustic energy to the system—it may not be considered as an ANC technique. Masking is a natural human auditive function: some signals can lift the hearing threshold levels at certain frequencies. With basis on this concept, the common idea of masking solutions attempts to mask high frequencies for making speech incomprehensible to adjacent rooms, for example, between a waiting room and a doctor's office, or between a meeting room and a reception area. In all these cases, the most common technique is music masking. Other options and applications are available as well, for example, the use of white noise or a water fountain sound for enhancing acoustic quality to the receivers [40], which has been studied for a long time to make soundscapes healthier [41].

5. Active noise control (ANC)

ANC refers to a technique that aims to reduce the SPL through the emission of sound signals in the counter phase with those that occur in the system to control. Then, destructive interference is generated between both signals. This method involves adding acoustic energy to a system that already has an excess of acoustic energy; thus, it should be applied very carefully. On the other hand, active silencers are more accurate and versatile than passive ones. ANC can be used in the source, during sound propagation or at the receiver. Best results should be expected when the signal is periodic, low frequency, and narrow band.

In a similar way to ANC techniques, there are active vibration control techniques. They act with analogous principles, but in this case, they introduce oscillations to counteract the original vibrations.

5.1 Main components of an active noise control system

An ANC system consists of the following components:

- A primary acoustic field, which is the acoustic field to be controlled.

- Detection sensors or transducers, which allow characterizing the signal to be controlled and provide the information to the system controller.

- Error sensors, which take information after the cancelation (that is, the sum of the primary signal plus the secondary signal -or control signal-), to inform the controller to perform the necessary corrections so that the result is as close as possible to zero.

- Actuators, which are electroacoustic devices (commonly, an amplifier and a loudspeaker) that receive the controller's instructions to emit a certain cancelation signal to modify the primary acoustic field ("anti-noise" or "counterwave", a sound with the same amplitude and frequency of the signal to control, but in opposite phase).

- A controller, on which the success of the system depends on. It is the electronic system that processes the signals from the error detection and control sensors, generates the signal to be emitted, and defines the delay with which it must be emitted; lastly, it sends the order to the actuators to emit it. The delay between the primary and secondary signal must be adjusted taking into account the primary detection sensors and the error sensors. The controller optimizes the attenuation of the error signal provided by an adaptive algorithm that varies the filter operating coefficients in real-time.

The system works as follows:

- A detection sensor (e.g., a microphone) receives the signal to be controlled and sends this information to the controller.

- The controller generates an equal wave (secondary field) but in the opposite phase and sends it to the actuator with a certain delay, which is the time that the controller calculates for the wave to reach the designated point.

- The actuator emits the signal indicated by the controller.

- Another sensor (an error sensor or error microphone) is located after the actuator receives the two added signals, primary and secondary, whose result should be as close as possible to zero, and sends it to the controller.

- As in practice, the result is not usually exactly null, the controller will make an adjustment to minimize the error.

5.2 Applicability of active noise control systems

The main rules of thumb for applying an ANC system are the following:

- The system should be as robust and powerful as possible.

- The minimum sampling frequency required by the controller must satisfy the Nyquist theorem: $f_{min,sampling} \geq 2f_{Max}$ of the frequencies to control. In other words, the minimum sampling frequency necessary to describe a signal of 500 Hz is at least 2 × 500, that is, 1000 samples per second. Hence, the lower the frequency of the signal to be treated, the less demanding this aspect becomes. Please, note that the Nyquist theorem is a necessary condition but not a sufficient one to achieve a good description of the signal through sampling. At first, this was a major limitation for the use of the ANC systems. Advances in communications technologies have broken down this barrier.

- The abovementioned condition implies that the processing time by the controller should be less than the time between two samples. The maximum available time between the income signal sent by the reference sensor and the emission of the cancelation signal by the actuator is usually a few milliseconds. For example, for 500 Hz, the minimum sampling rate is 1000 samples per second; thus, the processing time should be less than 1/1000 = 1 ms.

- The causality condition establishes the minimum length that the noise-canceling system must be above a certain frequency, to guarantee that the acoustic delay δA is being considered. To ensure that the adaptive filter has a causal response, it is necessary that $\delta A > \delta E$, δE is the total electrical delay (sum of the delays of the entire analog-digital system, including the controller, the digital filter, the loudspeaker, the processing time and so on. δA is the time taken by the flow to pass through the input microphone.

- The closer the two sources are (primary and secondary), the greater the reduction in the total sound power of the ensemble.

- The greatest efficiency of the solution occurs when the separation between sources is not greater than 0.1 λ: reductions of 10 dB or more can be achieved.

- If the separation between sources becomes greater than 0.3 λ, the attenuation will be negative, that is, the net acoustic power of the ensemble will increase.

- From the previous conditions, it turns out that ANC is more favorable the greater the wavelength of the signal to be controlled. These systems are generally applied for frequencies up to 500 Hz ($\lambda_{500\ Hz} \approx 0.68$ m).

- If the cancelation signal is emitted by more than one source, the greater the number of secondary sources, the better the result.

- To control a signal with significant amounts of energy in low and high frequencies, mixed solutions (active and passive) should be considered.

5.3 Control strategies in active noise control systems

There are two basic control strategies in ANC systems: anticipatory or feedforward control and feedback control.

The applicability of *feedforward control* relates to cases whether the propagation time between the reference microphone (the primary sensor) and the cancelation loudspeaker (the secondary source) is sufficient—in electrical terms—so that the canceling signal could be reintroduced downstream when the primary noise signal "passes" the loudspeaker. This is why active control was originally reserved for low-frequency sounds since at lower frequencies there was more time (or distance traveled) to insert the secondary signal. The objective of the system is to minimize the error signal so that the residual acoustic noise error (i.e., the sum of the original and input signals) results as low as possible.

The *feedback system* is simpler than the feedforward one, but it can only be applied if the signal to be controlled is well known and very stable, that is, with very few variations. Thus, the need to know the signal to be controlled at all times is avoided since the system assumes it is a fixed and known input.

A single microphone is used as an error microphone, which registers the sum of the signal to be canceled plus the cancelation signal. The information is sent to a simpler control system, which sends the signal to be emitted to the actuator aiming to minimize the error between the current signal and the predefined one.

A combination of both systems is also possible to be used ("hybrid control"). *Hybrid control* (FF/FB) refers to a two-degree-of-freedom controller, which enables robust performance over a larger range of frequencies. The design of each controller depends on the design of the other, for example, DEUS control structure output (two inputs, one output). In 2020, another structure for ANC systems was proposed [42]; it modifies the path for measuring the error.

Adaptive controllers may be mentioned as well. They allow good performance to be achieved even when the information of the signal to be processed is incomplete. When working under stationary conditions, an adaptive controller quickly converges to Wiener's optimal solution; when working in nonstationary conditions, the algorithm allows tracking time variations in the input data, as long as these variations are slow enough to track. Controllers can apply many possible algorithms, including the classic "least mean squares."

5.4 Some applications of active control

Possibly the most successful case in using ANC is that of active hearing protectors, which generate a counterwave to protect workers exposed to high SPL. ANC-ear protectors were one of the first applications of massive use of ANC; they met great commercial interest. Its use for the protection of workers is not the only application: not considering its dangers, it may be said that there are also motorcycle helmets with ANC, due to the occupational exposure of those who work on motorcycles (deliverymen, police officers [43]). Other nonpassive protectors are those that allow selective filtering of some frequency bands, such as those which are especially of interest to musicians; and those that allow masking the source by emitting a different signal (e.g., white noise, or even turning a radio on).

The applications of active control in the propagation medium are diverse and some of them are novel. A modern system for indoor urban noise control is through ANC. A 10 dB reduction in traffic and railway noise was achieved on frequencies from 100 Hz to 1000 Hz for canceling the incoming signals in a real window of 1 m $\times$ 1 m [44]. There are ANC systems for controlling SPL in the interior of a vehicle [45], a "noise-canceling office chair" [46] and a "capsule" for home devices, to control both noise and vibrations [47].

An interesting application of ANC is that proposed for a real case of thermoacoustic instabilities in large combustion engines. Shortly explained, thermoacoustic instabilities appear when working with machinery and equipment on such a scale that the main phenomena, such as mixing or ignition, cannot be considered punctual neither instantaneous nor homogeneous throughout the combustion

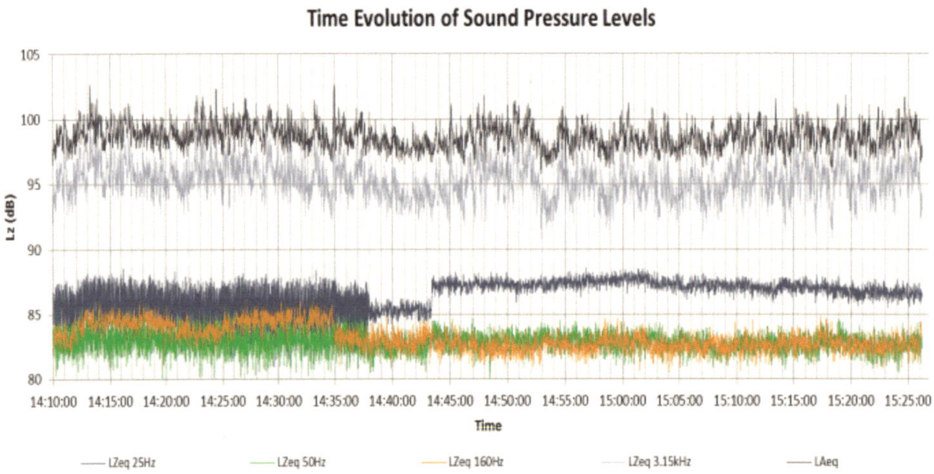

Figure 7.
Installation of thermoacoustic instability in the 25 Hz third-band octave from a large engine (from Ref. [49]).

chamber. Thus, the system can frequently enter into resonance if there is coupling with some geometric dimension.

When preparing this entry into resonance, a process of "loss of chaos" is observed [48], which reaches the condition of thermoacoustic instability from which, after some time that is not fixed or preestablished, the system recovers and returns to normal.

In this case study, a detailed analysis of the problem showed that fluctuations in the pressure of the fuel inlet line were occurring, leading to the occurrence of the abovementioned instabilities. One of the three ways in which these were manifested was by the loss of chaos in sound emissions, as described in Ref. [48]. The phenomenon appears especially in the 25 Hz third-octave band (blue line in **Figure 7**). The transition lasts several minutes, and it can be seen that the loss of chaos is the preamble for the emission in that band to present an abrupt increase of SPL about 5 min later. The suggested solution was a typical feedforward ANC system since there was a reaction gap of several minutes.

6. Final remarks

An overview of techniques for noise control has been presented. The most common passive solutions and their applications have been described; generalities and actuality of ANC have been reported as well.

This chapter will fulfill its objective if it serves only as a guide for selecting options for solving practical situations, but of course, a more in-depth approach is needed for their acoustic design.

Conflict of interest

The author declares no conflict of interest.

Author details

Alice Elizabeth González
IMFIA, Faculty of Engineering, Universidad de la República UdelaR (University of the Republic), Montevideo, Uruguay

*Address all correspondence to: elizabet@fing.edu.uy

References

[1] ASHRAE. Generator Noise Control—An Overview. USA: ASHRAE Technical Committee 2.6 on Sound & Vibration; nd

[2] Strahle WC. Combustion noise. USA: Journal of Progress in Energy and Combustion Science. 1978;**4**(3-A): 157-176

[3] Harris C. Manual de Medidas Acústicas y Control de Ruido. Mexico: McGraw-Hill; 1995. p. 1328

[4] Briggs J, Sacks M, Kraemer S. Heat Recovery Steam Generator Noise Attenuation. nd. 1995. p. 13

[5] Pettersson H. Environmental Noise Control in the Pulp and Paper Industry. Sevilla, España: Tecniacústica 2002; 2002. p. 6

[6] Gerges SNY. Ruido, Fundamentos e Controle. Brasil: Biblioteca Universitária, Universidade Federal de Santa Catarina; 1992. p. 600

[7] Pérez Barber D. Proyecto Fin de Carrera: Modelado acústico de silenciadores. España: Escuela Superior de Ingenieros de Sevilla, U. de Sevilla; 1992

[8] Giraldo Arango JM. Control del ruido en la construcción de obras públicas. In: Proceedings of International Seminar on Environmental Noise; March 2008; Medellín, Colombia. 2008 ISBN: 978–958–44-3029-8

[9] González AE. Noise sources in the city: Characterization and management trends. In: Siano D, editor. Noise Control, Reduction and Cancellation Solutions in Engineering. London, USA: IntechOpen; 2012. DOI: 10.5772/25879

[10] Teti L, de León G, Del Pizzo A, Moro A, Bianco F, Fredianelli L, et al. Modelling the acoustic performance of newly laid low-noise pavements. Construction and Building Materials. 2020;**247**:118509. p. 11

[11] Licitra G, Moro A, Teti L, Del Pizzo A, Bianco F. Modelling of acoustic ageing of rubberized pavements. Applied Acoustics. 2019;**146**:237-245

[12] Gruber M. Airfoil noise reduction by edge treatments [PhD thesis]. University of Southampton, Institute of Sound and Vibration Research, Faculty of Engineering and the Environment; 2012. p. 229

[13] Gianoli P, Cataldo J, González AE. Impacto acústico de la interacción entre el viento y las edificaciones. In: Revista do IPT do III Congresso Latino Americano de Engenharia do Vento; Sao Pablo, Brasil. 2018. p. 6

[14] Lisboa MR, Cataldo Ottieri J, González AE. Noise annoyance due to wind flow interaction with a building's facade. Open Journal of Acoustics. 2015; **5**:1-10. DOI: 10.4236/oja.2015.51001

[15] Government of the Hong Kong, Environmental Protection Department, Highways Department. Guidelines on Design of Noise Barriers. 2nd Issue. January 2003

[16] Peiró-Torres MP, Bravo JM, Redondo J, Ferri M, Sánchez-Pérez JV. Pantallas acústicas abiertas. In: Cuadernos tecnológicos de la PTC N°6/2015. Madrid, España: Plataforma Tecnológica Española de la Carretera (PTC); 2015

[17] European Commission. Directive 2002/49/EC and its modifications

[18] Yagüe DC, María J. Caracterización morfológica y geométrica de curvas tipo para diseño de pantallas acústicas

tubulares y su aplicación al campo de la protección ambiental en Ingeniería Civil [Tesis Doctoral]. Madrid, España: Universidad Politécnica de Madrid; 2001

[19] Yang T, Hu L, Xiong X, Petrů M, Noman M, Mishra R, et al. Sound absorption properties of natural fibers: A review. Sustainability. 2020;**12**:8477. DOI: 10.3390/su12208477

[20] Kotzen B, English C. Environmental Noise Barriers: A Guide to their Acoustic and Visual Design. 2nd ed. Canadá and USA: Simoultaneouly published; 2009. ISBN 10: 0-203-93138-6

[21] Kamisiński T, Kinasz R, Szeląg A, Rubacha J, Pilch A, Flach A, et al. The comprehensive research of the road acoustic screen with absorbing and diffusing surface. Archives of Acoustics. 2015;**40**(1):137-144

[22] Government of the Hong Kong SAR, Highways Department. Practice Notes No. BSTR/PN/003—Revision E. Noise Barriers with Transparent Panels. Bridges and Structures Division, October 2020

[23] Kolarević M, Grković V, Vukićević M, Bjelić M. Configuration of modular noise barriers. In: 4th International Conference on Noise and Vibration; Croacia. 2012

[24] Morgan SM, Kay DH, Bodapati SN. Study of noise barrier life-cycle costing. Journal of Transportation Engineering. 2001;**127**(3):230-236

[25] Guarnaccia C, Quartieri J, Mastorakis NE. Comparison of acoustic barriers noise reductions evaluated by different calculation methods. In: Latest Trends on Systems—Volume II. 2001. pp. 443-449. ISBN: 978-1-61804-244-6

[26] Pita Olalla E. Las pantallas antirruido y sus prestaciones. Revista de Obras Públicas. 1995;**3339**:77-88

[27] Ross JC, Arnoldy M, Evans J. New Hampshire department of transportation statewide noise barrier study. Transportation Research Record. 2018; **2672**(24):144-153. DOI: 10.1177/0361198118797188

[28] Zhaomeng W, Meng LK, Priyadarshinee P, Pueh LH. Applications of noise barriers with a slanted flat-tip jagged cantilever for noise attenuation on a construction site. Journal of Vibration and Control. 2018;**24**(22): 5225-5232. DOI: 10.1177/1077546317747779

[29] Government of the Hong Kong Special Administrative Region, Greening and Landscape Office, Greening, Landscape and Tree Management Section. Guidelines on Greening of Noise Barriers. 2012. p. 20

[30] Lee SE, Velasquez S, Flintsch G, Peterson J. Road noise attenuation study: Traffic noise, trees, and quiet pavement: A report in response to item 442 c. of Chapter 847 of the Acts of Assembly of 2007. The Virginia Transportation Research Council; 2008

[31] Ruza Tarrío Felipe. La vegetación en la lucha contra el ruido. 2008. p. 10

[32] Van Renterghem T. Guidelines for optimizing road traffic noise shielding by non-deep tree belts. Ecological Engineering. 2014;**69**:276-286 277. DOI: 10.1016/j.ecoleng.2014.04.029

[33] Horoshenkov KV, Khan A, Benkreira H. Acoustic properties of low growing plants. JASA. 2013;**133**:2554. DOI: 10.1121/1.4798671

[34] Asdrubali F, D'Alessandro F, Mencarelli N, Horoshenkov KV. Sound absorption properties of tropical plants for indoor applications. In: The 21st

International Congress on Sound and Vibration; July 2014; Beijing/China. 2014

[35] Rizzo La Malfa A, Velis A, Posse C, Armas A, Miguel S, Figueira A, et al. Sistemas modulares verticales: Evaluación de la absorción acústica en laboratorio. Revista Ecos. 2020;**1**(1):48-57

[36] Azkorra Z, Pérez G, Coma J, Cabeza LF, Bures S, Álvaro JE, et al. Evaluation of green walls as a passive acoustic insulation system for buildings. Applied Acoustics. 2015;**89**:46-56

[37] Redondo J, Sánchez-Pérez JV, Barreiro J, Iluminada P. Sobre el rendimiento de difusores de sonido basados en cristales de sonido. In: 45° Congreso Español de Acústica, 8° Congreso Ibérico de Acústica, European Symposium on Smart Cities and Environmental Acoustics; Murcia, España. 2014. pp. 787-793

[38] David RS. Utilización del método de los elementos finitos para el diseño teórico de una pantalla acústica basada en cristales de sonido [Graduation thesis]. Escuela Técnica Superior de Ingenieros de Telecomunicación, Universidad Politécnica de Valencia; 2019

[39] Rubio C, Castiñeira-Ibáñez S, Sánchez-Pérez JV, Candelas P, Belmar F, Uris A. Chapter 6: Open acoustic barriers: A new attenuation mechanism. In: Advances in Noise Analysis, Mitigation and Control. Croatia: IntechOpen; 2015. pp. 141-159. DOI: 10.5772/64360

[40] Piana EA, Roozen NB. On the control of low-frequency audible noise from electrical substations: A case study. Applied Science. 2020;**10**(637):20. DOI: 10.3390/app10020637

[41] Kogan P. Acústica Ambiental y el Paradigma del Paisaje Sonoro: Investigación exploratoria en áreas verdes y otros espacios urbanos [Tesis Doctoral]. Córdoba, Argentina: Universidad Nacional de Córdoba; 2018 . p. 274

[42] Figwer J, Michalczyk MI. Notes on a new structure of active noise control systems. Applied Sciences. 2020;**10**: 4705. 20 pp

[43] Rosa C, Peña S, Ricardo S. Control activo de ruido acústico en cascos de motociclismo. Revista Iberoamericana de Automática e Informática Industrial. 2007;**4**(3):73-85

[44] Lam B, Shi D, Gan W, Elliott SJ, Nishimura M. Active control of broadband sound through the open aperture of a fullsized domestic window. Nature Research, Scientific Reports. 2020;**10**:10021

[45] Lee S, Lee S, Back J, Shin T. A new method for active cancellation of engine order noise in a passenger car. Applied Sciences. 2018;**8**:1394

[46] Sujbert L, Szarvas A. Noise-canceling office chair with multiple reference microphones. Applied Sciences. 2018;**8**: 1702. DOI: 10.3390/app8091702

[47] Mazur K, Wrona S, Pawelczyk M. Performance evaluation of active noise control for a real device casing. Applied Sciences. 2020;**10**:377

[48] Vineeth N, Gireehkumaran T, Sulochana K, Saravanan G, Sujith R. Loss of chaos in combustion noise as a precursor of impending combustion instability. International Journal of Spray and Combustion Dynamics. 2013;**5**(4):273-290

[49] González AE, Cataldo-Ottieri J, Gianoli-Kovar P, Montero-Croucciée J, Lisboa MR. Measuring sound pressure levels during thermoacoustic instabilities in large engines: Case study. Journal of Modern Physics. 2017;**8**:1685-1699. DOI: 10.4236/jmp.2017.810099

Chapter 3

Approaches for Noise Barrier Effectiveness Evaluation Based on *In Situ* "Insertion Loss" Determination

Antonio Barba and Juan M. Martinez-Orozco

Abstract

In situ evaluation of the effectiveness of noise barriers may be based on the assessment of their intrinsic or extrinsic characteristics. The evaluation of intrinsic characteristics is based on acoustic properties, such as noise barrier absorption or insulation. The evaluation of the extrinsic characteristics is based on the calculation of the barrier Insertion Loss, which is defined as the difference in the noise level before and after the installation of the barrier. Insertion Loss is calculated using two different approaches: the direct and indirect methods. The direct method is used when the barrier has not been installed yet or can be removed, while the indirect method is used when the barrier is already installed and cannot be easily removed. This chapter describes the different approaches used in the scientific literature for *in situ* evaluation of the effectiveness of noise barriers and discusses the noise attenuation levels obtained with each approach.

Keywords: diffraction, effectiveness, Insertion Loss, noise barriers, traffic noise

1. Introduction

1.1 Noise barriers: the ubiquitous solution to the road noise problem

Chronic exposure to environmental noise is a widespread problem around the world, causing significant impacts on human health and well-being. Road traffic is the predominant source of noise in urban areas and represents the second most important health risk factor after air pollution [1]. In Europe, it is estimated that about 20% of the total population is exposed to road traffic noise levels considered harmful to health [2]. Moreover, the problem is expected to become more severe in the next decades. In the European Union, the population exposed to high road noise levels is projected to rise both inside and outside urban areas over the next years due to urban growth and increased demand for mobility [3].

In the last decades, the introduction of more stringent environmental noise legislation has resulted in a series of noise abatement measures of varied nature. These included urban planning measures (such as the designation of noise-sensitive areas, or regulations on vehicle speed limits or traffic restrictions), measures to improve the acoustic performance of vehicles, pavements and buildings, and the construction of noise barriers. Currently, noise barriers have become frequent features along many roads and railways.

The history of noise barriers precedes the appearance of the first generation of environmental regulations in the World. The first documented noise barrier installed on a road was built in 1963 [4]. In the following years, several new design criteria and new materials for barriers were rapidly introduced. And so, the first lightweight barriers with an absorptive treatment on the panel surface date back to the early 1970s [5]. By 1975, Japan had already built noise barriers along 79 km of new highways [6], and the USA had installed about 57 km of barriers at certain types of highway projects [4].

In the 1960s and 1970s, the first research studies were initiated to analyze the acoustic properties of barriers and calculate noise attenuation levels. Probably the most famous of these studies was the Maekawa empirical chart of 1968 [7], as well as the formulations developed by other authors based on Maekawa's original proposal [8–12]. It was also during this period that the first regulations for noise management and abatement were adopted in countries such as the USA (1972), Canada (1973), Germany (1974) and Japan (1974).

Since the 1980s, many countries have adopted Environmental Impact Assessment (EIA) legislation that requires the evaluation of, among others, proposed road projects that are likely to have significant environmental impacts. As part of the EIA process, the project developer is required to evaluate road traffic noise and must determine appropriate mitigation measures to minimize its effects. Constructing a noise barrier is probably the most mentioned mitigation measure in EIAs conducted around the world [13]. As an example of the extensive use of these devices, it was estimated that the global production of noise barriers reached approximately 370 million m^2 in 2014 [14]. In the European Union, these devices have become the most prominent noise mitigation measure applied to major roads located outside residential areas [15]; in the USA, about 5700 km of barriers have been built to date [4].

Noise barriers have been made of many different materials and have taken many different forms over time. In the past, simple reflecting barriers made of concrete, masonry blocks, or earth berms were often used, but modern barriers tend to have absorptive treatments which minimize the level of reflected noise. In recent years, a number of innovative barriers are being developed, such as combined noise and safety barriers, low-height barriers, photovoltaic barriers, noise walls with titanium dioxide (TiO_2) coating, inox/corten steel barriers, or acoustic devices based on sonic crystals [16].

The growth in the use of noise barriers has also been coupled with a growing interest in their effectiveness as a tool to reduce noise pollution. The evaluation of this effectiveness is, however, a difficult task, given that these devices are placed outdoors under very varied conditions, with diverse barrier designs and locations, fluctuating noise sources, and changing environmental conditions. This chapter outlines the fundamentals of the acoustic performance of barriers, describes the main approaches for the evaluation of their effectiveness, as well as the main findings obtained in the studies conducted on the attenuation levels measured.

1.2 Acoustic performance of noise barriers

A noise barrier is a structure that obstructs the direct transmission of airborne noise produced by a source, such as road traffic, and redistributes the sound energy into several paths (**Figure 1**):

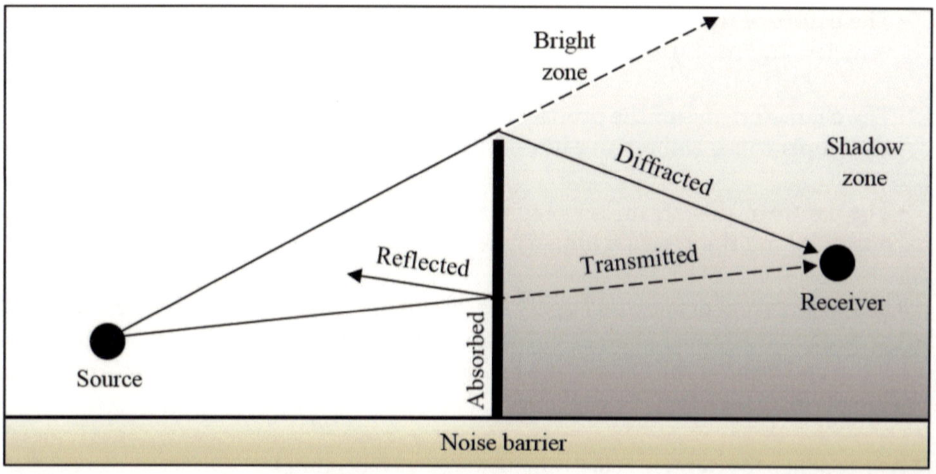

Figure 1.
The acoustic performance of a noise barrier (based on [17]).

- A reflected path, so that the noise wave reaching the exposed side of the barrier partly reflects on it. The barrier can also absorb other parts of the sound energy. Based on these acoustic properties, noise barriers are usually divided into two main groups: absorptive barriers, which are specifically designed to absorb part of the acoustic energy, and reflective barriers, from which noise is largely reflected (a special group consists of reactive barriers, which are devices that contain cavities or resonators).

- A transmitted path, so that the noise reaching the exposed side of the barrier transmits through the device itself. Therefore, the transmitted energy must be as low as possible.

- A diffracted path, over the top and around the ends of the barrier, so that the barrier acts as an obstacle to the noise propagation, diffracts noise waves, and then propagates to the protected side of the barrier with less energy. Noise diffraction is largely determined by the difference between the source-receiver direct path length and the extended path length due to the presence of the barrier.

Noise barriers cause an area of decreased sound energy behind the barrier (also called shadow zone) which is a combination of reflection, diffraction, and transmission losses. Due to the nature of sound, diffraction does not bend all frequencies uniformly: higher frequencies are diffracted to a lesser degree; lower frequencies are, by contrast, diffracted deeper into the shadow zone behind the barrier. As a result, noise barriers are generally more effective in attenuating the higher frequencies.

The acoustic performance of a noise barrier depends on a set of intrinsic and extrinsic characteristics [18]. Intrinsic characteristics refer to the properties of individual components of the barrier, such as the type, thickness, and design of the materials used. Extrinsic characteristics consider the attenuation of the barrier once it has been installed. These characteristics are mainly determined by a set of context-specific conditions, such as:

- The position of the barrier relative to the source and the receiver, and its effective height and length to block propagation paths.

- The nature of the noise source in terms of traffic volume, traffic speed, types of vehicles, and road pavement.

- The characteristics of the propagation medium, i.e., wind conditions, air temperature, and relative humidity.

- The nature of the terrain between the road and the receiver, i.e., interfering obstacles and the acoustic impedance of the ground surface.

These contextual properties largely determine the diffraction characteristics of the barrier and the global noise attenuation that can be achieved. The noise diffracted on the top and around the ends of the barrier is the most important factor limiting its acoustic performance [18].

1.3 Approaches for determining noise barrier effectiveness

Determining the effectiveness of noise barriers has attracted the attention of researchers for the past 40 years, and a wide variety of both mathematical and experimental approaches have been developed. Mathematical methods have been widely used to determine the diffraction properties of the barriers. These methods can be based on the boundary element method [19, 20], the finite element method [21, 22], and the finite difference method [23].

Experimental studies have been based on diverse approaches relating to (i) the assessment of perceived annoyance reduction efficiency of noise barriers [24, 25], (ii) the effects of noise barriers on the perception of urban soundscape quality [26], (iii) the measurement of noise attenuation based on scale model experiments [27, 28], and (iv) the measurement of the acoustic properties of full-scale barriers. The latter experiments have been the most reported in the literature, and have addressed the analysis of the effectiveness of barriers based on their various acoustic characteristics:

- Some research studies have addressed the intrinsic characteristics of barriers, such as sound absorption and insulation. Two types of measurement methods are commonly used to evaluate these properties: laboratory methods, using a diffuse sound field in a reverberation room, and *in situ* methods.

- Other studies assessed barrier performance by measuring its "Insertion Loss", which is defined as the difference in sound pressure level before and after the barrier is constructed.

The methods for *in situ* evaluation of barrier effectiveness are described below, with a particular emphasis on those based on the determination of Insertion Loss.

2. In situ tests for evaluating intrinsic characteristics of noise barriers

A substantial part of the scientific literature on the evaluation of the acoustic properties of noise barriers has been based on the *in situ* evaluation of their intrinsic characteristics.

Figure 2.
Measurement according to the "Adrienne" method (left) [30] and "QUIESST" method (right) [31].

In Europe, *in situ* tests for measuring intrinsic characteristics have been based on standardized methods such as AFNOR 31089 [29], the European projects "Adrienne" [30] and "QUIESST" [31], and more recently the EN 1793-4, 5 and 6 standards [32–34].

The measurement system consists of a fixed source (loudspeaker) reproducing a maximum length sequence (MLS) signal [30, 31] or a gunshot [29]. With these kinds of signals, the impulse response of an acoustic system can be obtained. In addition, background noise is eliminated [35]. Then a microphone is located behind the barrier to measure noise transmission or/and in front of the screen to measure noise reflection (**Figure 2**).

These methods are focused on the measurement at the near field (placing the microphones close to the surface to be measured) since, according to the standards, the lower power of the waves reflected, the difficulty of discerning between emitted and reflected sound, and the influence of background noise, make it really difficult to obtain meaningful results more distance [30, 31]. For this reason, some researchers prefer to extrapolate reflectivity data measured in the near field toward the effect in the far-field.

However, other researchers based on the standard EN 1793-4 [32], where receiver microphones are placed 2 m behind the barrier (see **Figure 3**), have situated the receiver microphones at greater distances from the screen (from 10 to 40 m) [36] considering these distances better to estimate the real IL.

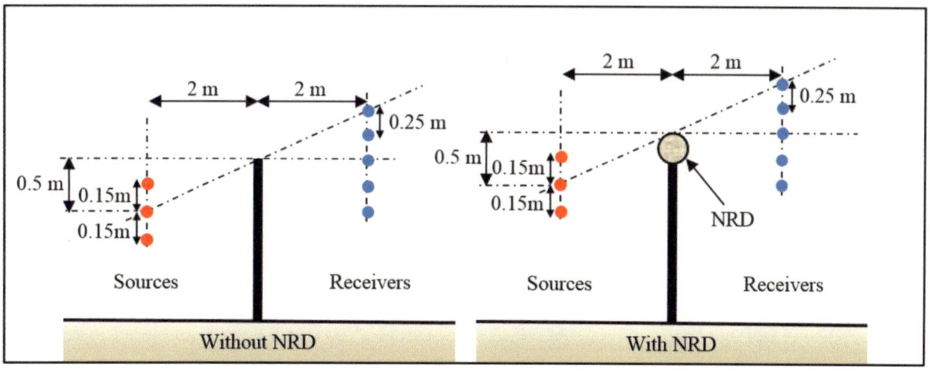

Figure 3.
Standard 1793-4 microphone location (based on [32]).

In addition, according to Kim et al. [36], since the European test methods are based on an impulsive signal, they could be eliminating the influences of reflected sounds by ground and any other objects around the test area. So that, methods such as conventional Japanese and Kim et al. use traffic signal for their measurements.

3. In situ tests for determining noise barrier Insertion Loss

Part of the research studies conducted to date has been based on the evaluation of the effectiveness of barriers on the calculation of Insertion Loss (IL), which is defined as the difference in the noise level before and after the installation of the barrier [37]. The IL is an extrinsic characteristic of noise barriers, depending mostly on the site geometry, meteorological conditions, ground impedance, and the relative positions of the noise source and the receiver [17]. These factors are in general not independent of each other, so the total IL cannot be calculated by the addition of partial insertion losses [17].

The international standard ISO 10847:1997 [37] establishes two methods of in situ IL measurement and calculation; direct and indirect measurement methods:

- The direct method is used when the barrier has not been installed yet or can be removed. The noise level is measured before and after the installation of the barrier to determine the IL. In this method, it must be ensured that measurements before and after the installation of the barrier are performed under equivalent weather and traffic conditions.

- The indirect method is used when the barrier is already installed and cannot be removed. In this case, an estimated "before" noise level is obtained by the measurement at a site that is considered equivalent to the study site.

The American standard ANSI/ASA S12.8-1998 [38] describes an additional "indirect predicted" method, which uses measurements at the site with a barrier to determine "after" noise levels, and a traffic noise prediction model to predict "before" levels at the same site without the barrier.

The ISO standard specifies general criteria for in-situ measurement of barrier IL including microphone positions, noise source conditions, and acoustic environments of the measurement sites. It also suggests generic principles for ensuring that sufficiently equivalent conditions are maintained between "before" and "after" measurements to permit reliable determination of barrier IL. The noise descriptor recommended is A-weighted equivalent sound pressure level. The materialization of the general criteria suggested by ISO has been resolved in different ways in studies based on both the direct and indirect methods.

3.1 Direct measurement method

The direct method described in the ISO 10847 standard is the approach to be used when the barrier has not been installed yet or can be removed. The method requires measurements before the barrier has been constructed to determine "before" levels and measurements at the same site after construction to determine "after" levels. According to US Federal Highway Administration (FHWA) [17, 39] this method ensures identical site geometric characteristics but also requires equivalent "before" and "after" meteorological and traffic conditions that may be difficult to reproduce. These meteorological equivalence conditions include wind, temperature, humidity, and cloud cover. In case of strong winds, "before" and "after" measurements should be avoided.

The factors to be considered in the determination of the measurement sites and procedures are briefly described below:

3.1.1 Noise emission source

The ISO standard recommends the traffic itself as the sound source for the "before" and "after" measurements. Using traffic noise signal has the advantage of measuring the signal which is wanted to evaluate. However, the fluctuations in traffic may affect the accuracy of the results, so that the measurement period must be taken into consideration.

3.1.2 Microphone locations

The standard recommends the use of a reference microphone, which allows for calibration of "before" and "after" measured levels and helps to consider variations in the characteristics of the noise source [17, 39].

When the reference microphone is used, it is placed in most cases according to the ISO standard, i.e., at a point on a vertical plane including the barrier, and at a height, at least, 1.5 m above the barrier edge. When the barrier is located less than 15 m from the near road lane, the microphone may be placed at 15 m from the center of the road lane, and at a height such that the line-of-sight angle between the microphone and barrier top, as measured from the center of the near road lane, is at least 10° (**Figure 4**) [17].

Microphone location at receiver positions depends on the study objectives since the location of the microphones (distance from the barrier, height above the ground) are determinants to establish diffraction effects. In some studies, a single microphone is located at a height of 1.5 m above the ground. The most common situation is, however, to place microphones at different distances and heights [17, 39–41] for a better understanding of the performance of the shadow zone (**Figure 5**).

3.1.3 Measurement period

A range from 2 to 30 min is the usual sampling period for measurements. When weather conditions are fluctuating, longer sampling periods, such as 1 or 24 h, could be more accurate [17]. It has been suggested [40] that 2-min measurements, as specified in the standard, are too short for a stable and reliable evaluation of sound pressure levels. The optimal measurement periods are 15–30 min, as longer periods

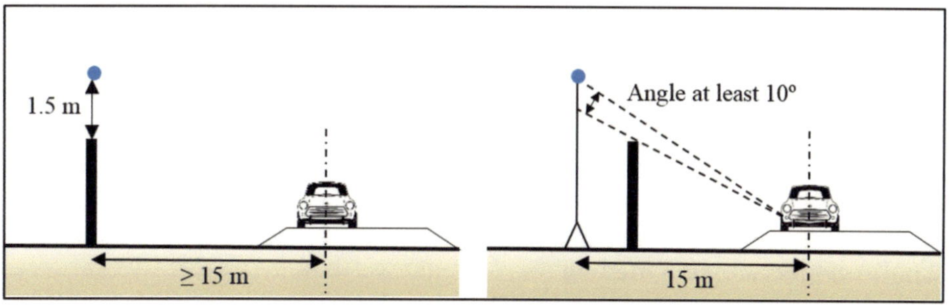

Figure 4.
Alternative positions for reference microphones—blue circle—(based on [17]).

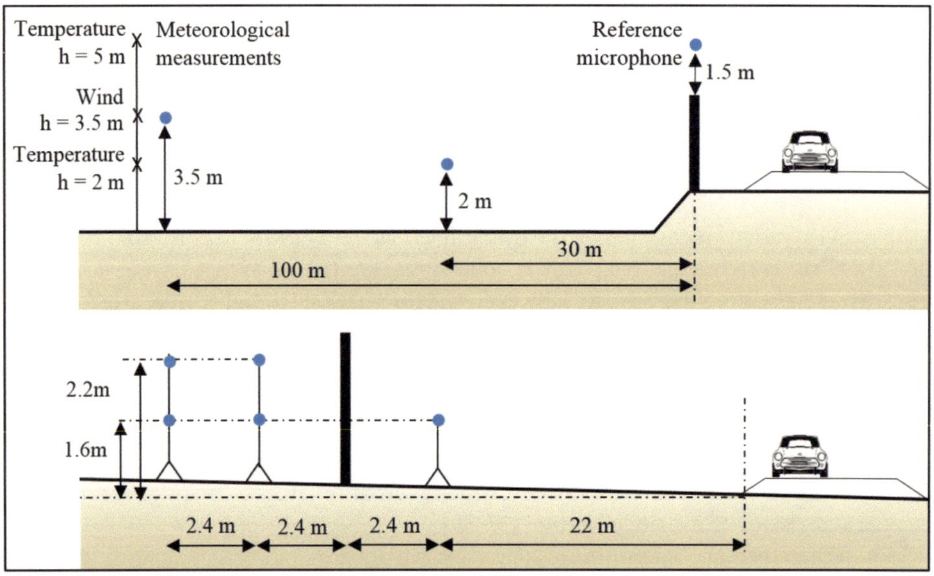

Figure 5.
Measurement microphones in the study of Anfosso-Lédée et al. [40] (above) and Parnell et al. [41] (below).

would possibly introduce atmospheric changes [40]. However, other researchers [41] established a 10 min measurement period.

The FHWA suggests avoiding measurements when wind speed exceeds 17 km/h or while raining since raindrops generate noise and tire noise increases on wet pavements. The Agency also recommends avoiding measurements when traffic flow is congested since the traffic noise level will be lowered, making it more difficult to evaluate the IL.

3.1.4 Main findings

There is a lack of evidence on the effectiveness of barriers based on the direct method. In the study conducted by Anfosso-Lédée et al. [40], the authors suggest that it could be due to the poor applicability of the method since it took 3 years for his team to complete the measurements due to the time waited for the installation of the barrier and the difficulty to find the equivalence of traffic, weather conditions, and ground impedance.

The experiment was based on measurements at 30 m and 100 m from the barrier (or where the barrier was supposed to be installed) (**Figure 5**). The measurements "before" took place in 1996, and the measurements "after" (when the barrier was already installed) were implemented from June 1998 to August 1999. Results showed IL values range from 4 to 8 dB(A) at 100 and 30 m from the road.

Parnell et al. [41] constructed a new barrier 80 m long and 2.4 m high and measured, with and without the barrier, at a distance of 2.4 m in front of the barrier and 2.4 and 4.8 m behind the barrier (**Figure 5**). Results of the experiment showed a 6–8 dB(A) difference in measurements "before" and "after" the barrier installation.

3.2 Indirect measurement method

The indirect method is, according to ISO 10847:1997, the approach to be used when the noise barrier has already been installed and cannot be removed for measurements.

In this case, an estimated "before" noise level is obtained by the measurement at a site that is considered equivalent to the study site. To ensure consistency of results, the "before" and "after" measurements should be performed simultaneously.

The indirect method is the only practicable approach in the case of most new roads, where the noise barriers have been installed during road construction, and therefore it is not possible to obtain a "before" measurement under normal traffic conditions. The primary advantage of using this method is that it ensures the same environmental conditions (meteorological and traffic conditions), so this method, as highlighted by some authors [17, 42], would be preferred over the direct measurement method.

The use of the indirect method involves the identification of another measurement site that is deemed to be equivalent. For these equivalent sites, a close match is required in emission characteristics, relative positions of source, barrier and receiver, acoustic performance of ground surface, terrain profile, interfering obstacles, reflecting surfaces, and meteorological conditions. The factors to be considered in the determination of the measurement sites and procedures are briefly described below.

3.2.1 Selection of equivalent sites

According to ISO 10847:1997, the "before" site must have a terrain profile, interfering obstacles, and reflecting surfaces equivalent to those of the real barrier site within a sector extending 60° on either side of the line connecting the receiver positions towards the source position, so that similar noise propagation can be achieved.

It is also necessary to ensure the equivalence of ground surface, which refers to the acoustic impedance of the ground along the source-receiver propagation path (i.e., acoustic characteristics of soil coverage, such as paved soil, vegetation on loose or packed soil, gravel, etc.) (**Figure 6**). The standard ISO additionally requires that

Figure 6.
An example of "after" and equivalent "before" locations at one of the sites studied by the authors in Spain (aerial photograph from Iberpix, OrtoPNOA 2020 CC-BY 4.0 scne.es).

the environment in the region within 30 m behind and to the side of the receiver positions shall be similar.

The main difficulty of the method is that an adjacent equivalent site may not always be available, especially in dense urban areas [17, 40]. As an example, a study conducted in Spain [42], which was based on an initial sample of 84 measurement sites, had to reject 54 potential locations due to various causes; the main cause was the different acoustic environment at the "before" and "after" positions due to significant differences in terrain profile and the presence of other noise sources.

3.2.2 Noise source

Most of the indirect method-based studies use road traffic as a noise source. The ISO standard proposes that naturally occurring road noise should be used as the sound source equivalence for the "before" and "after" measurements. The use of traffic noise has the obvious advantage of representing the natural source, but also the disadvantage of describing fluctuations in traffic volume, speed, and composition that may affect the accuracy of the results.

The use of artificial noise sources is infrequent and is used only when it is not possible to use traffic noise in equivalent conditions. This artificial point source may be based on a loudspeaker that reproduces traffic noise [20] or a regulated artificial signal such as pink noise [43].

3.2.3 Microphone locations

One of the key factors in the use of the indirect method is that the locations of the microphones relative to the noise source at the "before" and "after" positions should be identical, in terms of distance from the road and height above the road [39]. Some authors suggest the use of a reference microphone [17, 39], which, as mentioned before (Section 3.1.2) takes into account the effect of possible fluctuations of the noise source.

Only a few studies have considered the use of the reference microphone [39, 44], so it is understood that the rest of the studies assume that possible traffic fluctuations during the measurements are not expected to significantly affect the results.

The location of the receiver microphones varies according to the purpose of the study. The choice of these locations is sometimes determined by the possibility of finding equivalent locations at the "before" site.

In most studies, microphones are placed at regular distances from the barrier (5, 10, 15 m), or corresponding to incremental doublings of the distance (e.g., 7.5, 15, 30 m) [42, 44]. Some studies determine IL levels by placing a single microphone in the near field behind the barrier, at distances of 1–5 m [20, 43, 45, 46]. The most common height for the microphone is 1.5 m, although there are studies that consider additional heights, which are similar to or higher than the barrier height (e.g., 2, 4, 6 m). Both the distances and incremental heights of the microphone positions are intended to better understand the performance of diffraction shadow zones (**Figure 7**).

There is no general standard for receiver locations. The ISO standard proposes general criteria that are a very general characterization of the open space behind the barrier [47]. In recent years, the European Committee for Standardization adopted the CEN/TS 16272-7:2015 standard for railway noise barriers [48], which recommends nine locations for receiver microphones. These microphones are located at a distance of 7.5, 12.5, and 25 m away from the lines, and at a height of 3.5, 6, and 9 m

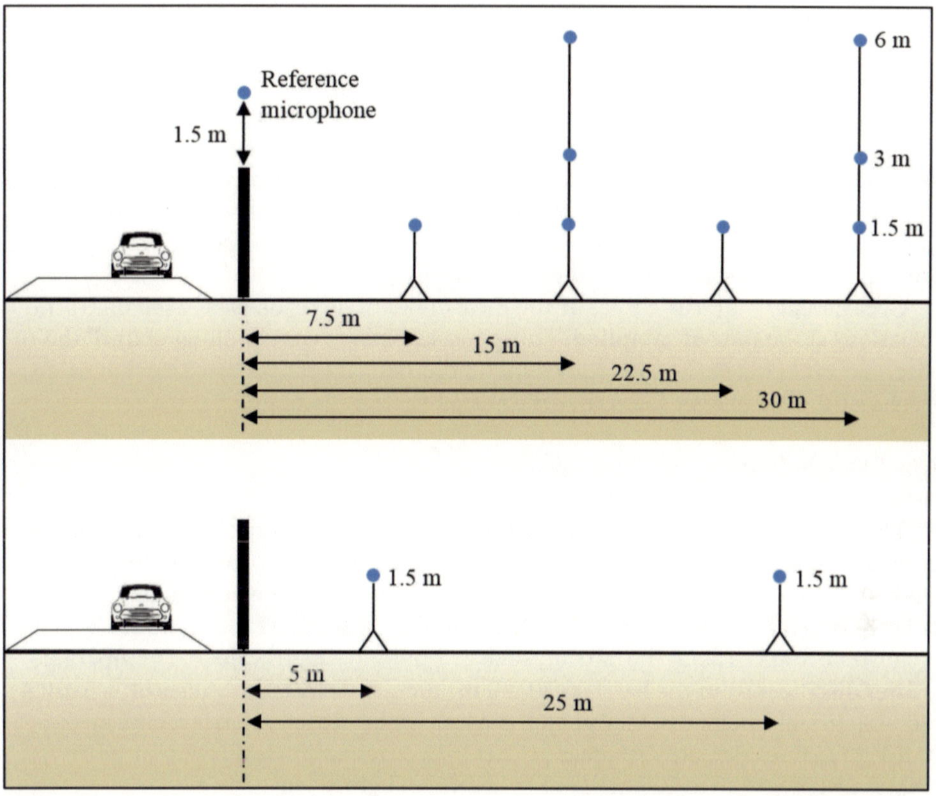

Figure 7.
The experimental design of studies based on the indirect method depends on the purpose of the study. Above, a microphone distribution is intended to better understand the pattern of the shadow zone [44]. Below, microphone distribution to measure IL levels at a distance at which real receivers are located (near the building facade) [42].

above the ground. However, this standard does not appear to be in use in studies relating to the measurement of Insertion Loss at railway noise barriers [47].

3.2.4 Measurement period

The selection of the measurement period should first consider when to measure along the daily time. One of the factors to be taken into consideration concerns favorable weather conditions, in particular wind speed and direction. The preferred conditions are for daily periods when low wind or calm is expected. Some studies [42] have conducted measurements in the period after peak traffic time in order to find dense but fluid traffic conditions, where traffic fluctuations are less prominent. In most of the studies, the "before" and "after" measurements have been undertaken simultaneously to ensure the same environmental conditions (i.e., background noise, traffic, and meteorological conditions).

The duration of the measurements in studies based on the indirect method depends on the nature of the noise source. In the case of studies using an equivalent artificial noise source, the duration of measurements is usually short (such as 2 min) in accordance with the ISO standard [43]. In the case of road traffic noise, the period is usually long enough to ensure the representativeness of the spectrum of the traffic noise. In practice, measurement duration in most studies ranges from 10 to 30 min, and the most common value is 15 min. Some studies [17] have suggested using longer periods (such as 1 h, or a day) when noise variations are expected to be substantial, but these longer periods do not seem to be used in practice.

In other studies [39, 49] the procedure consists of measuring noise levels, wind speed and direction, and temperature lapse rate for a 4-h block of time in 1-min increments. Thus, the results are broken down into short periods and continuous equivalent levels and meteorological conditions are individually determined for each short period. This procedure anticipates the problem of *a priori* considering possible fluctuations in the meteorological conditions of the site.

Occasionally [42], the choice of the measurement duration was based on traffic variations at the time of sampling. Thus, measurements were prolonged until the observed variation in the sound level meter did not vary more than a certain value (such as 0.1 dB(A)) over a certain time period (at least 1–2 min).

3.2.5 Main findings

The results obtained in the different research studies conducted revealed moderate Insertion Loss values of the noise barriers. Attenuation values obtained in the near field, at distances from the barrier of 5–7 m, and heights above ground of 1.2–1.5 m, range between 7 and 10 dB(A) [20, 42, 44–46, 50]. Insertion Loss levels are higher at shorter distances from the barrier, such as 1 m [43]. The IL values at comparable greater distances from the barrier (20–30 m) tend to decrease to values of 3–5 dB(A) [40, 42, 44], although one study reports much higher attenuation levels of up to 10 dB(A) at intermediate distances (15 m) [51]. Attenuation levels measured at greater distances (up to 100 m) tend to decrease slightly [40].

These results seem to indicate that the barrier attenuation levels are, above a certain distance, clearly lower than expected. It is, however, generally assumed that an effective noise barrier typically reduces noise levels by about 5–10 dB(A) [16, 44]. Effectiveness usually depends on its dimensions, material type, and location relative to the source and receiver positions. In the dimensioning of the barrier, the contribution to the total sound field of the components diffracted around the top and side edges are the key elements to determine the minimum barrier height/length for which the influence of the side edges diffraction may be neglected.

The best noise reduction effect is in the frequency range of 250–4000 Hz, at which the traffic noise is dominant. The average value of Insertion Loss for the octave bands between 250 Hz and 4 kHz ranges from 4 to 9 dB(A) [42, 50]. Noise abatement reaches a maximum at 4000 Hz, and the smallest reductions are encountered for the lowest frequencies (**Figure 8**) [42, 50, 53].

The type of barrier material does not appear to have a significant effect on attenuation levels [42, 51]. The differences found are rather related to locational factors, such as the distance from the barrier to the source (or receiver). Thus, the Insertion Loss measured at earth berms is lower than at noise walls because the top edge of the barrier is usually further away from the source and/or receiver positions.

3.3 Equivalence of direct and indirect methods

There is little evidence of equivalence of the results obtained with the direct and indirect methods. In the only study conducted to date evaluating the IL of the same site (the same noise barrier) using both methods [40], the results reveal that the direct and indirect methods are not equivalent. The observed differences range from

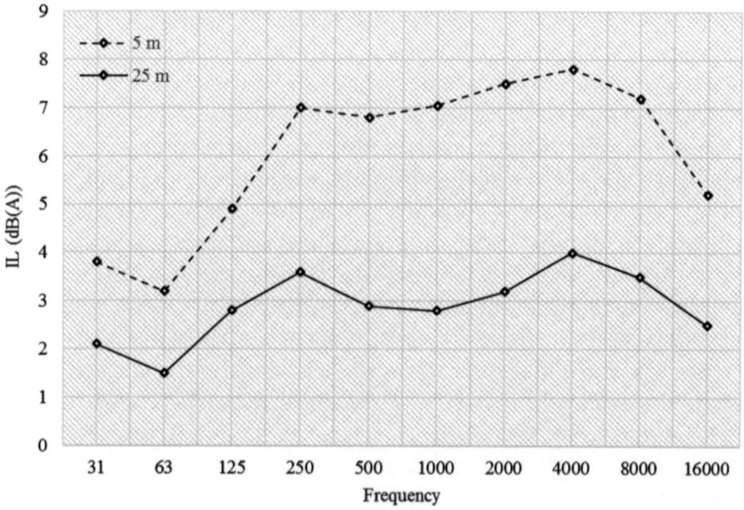

Figure 8.
An example of Insertion Loss levels in the range of frequencies of the octave bands at two distances (5 and 25 m) from the noise barrier [52].

−2 dB(A) to +4 dB(A). The causes of these differences were attributed in the study to variations in wind conditions (wind speed and direction) and vertical temperature gradient. The effect of microphone positions and other environmental factors on noise levels measurements also needs to be better known.

4. Conclusions

The amount of literature on the effectiveness of noise barriers has not provided sufficient evidence on the actual attenuation achieved by these devices, and there is uncertainty over the noise reduction capabilities of existing barriers. The methods described in the ISO 10847:1997 standard have drawbacks that make it difficult to obtain reliable attenuation measurements.

The direct method ensures identical propagation characteristics since the source of noise, the barrier, and the receiver are at the same positions, but the equivalence of source and meteorological conditions may not be fully satisfied. The indirect method ensures that the same local weather and traffic conditions are maintained, but the equivalence of terrain profiles, obstacles, and ground surface conditions may not be fully achieved. In addition, the usage of provisions of the ISO standard sometimes is complicated when at the site point exists a relatively high background noise level, or adverse meteorological conditions [54].

According to the ISO standard, the recommended method is the direct method, although most studies have been based on the indirect method because the barriers were installed during road construction, and therefore it was not possible to obtain equivalent "before" measurements.

The ISO standard provides generic methods for determining Insertion Loss at receiver locations. However, there are no universally acknowledged receiver positions for measurements. It is important to note that barriers are relatively ineffective at some distance from the road. The effective distance range is limited to a few tens of meters, so it is unclear that many receivers can benefit from barrier attenuation. Many of the studies conducted have calculated Insertion Loss levels at barrier near-field distances, so the noise reduction capabilities of barriers were only partially assessed. The IL levels measured at comparable greater distances from the barrier (20–30 m) were, in most cases, very moderate. This supports the argument that the barrier attenuation levels are, above a certain distance, clearly lower than expected.

Additionally, the ISO standard specifies measurements of equivalent continuous A-weighted sound pressure levels to calculate the attenuation of the barrier. However, A-weighting tends to underestimate the effects of low-frequency noise [47]. Several studies have highlighted that A-weighting does not adequately consider the perceived annoyance produced by predominantly low-frequency noise. This is the case of road traffic noise, which is characterized by the wide variability in the relative level of low-frequency noise [25, 55]. Noise barriers increase the relative level of low-frequency of noise on the shielded side of the barrier. Thus, the attenuation in A-weighted measured levels level may overestimate the estimated reduction in perceived annoyance due to the increase in the relative level of low-frequency sound [47].

In summary, the literature has described some critical points about the applicability and reliability of ISO methods. These points were dealing with (i) the reliability of results (i.e., direct IL measurements obtained at different moments, and indirect IL measurements obtained at equivalent locations), (ii) the equivalence of results of direct and indirect methods, (iii) the nature of the indicator used (A-weighted levels), and (vi) the relevance of operational factors such as weather conditions, traffic fluctuations, ground impedance, and background noise.

The effect of these factors on noise levels measurements needs to be better known. More research studies in this domain are required to bring improvements in measurement methods.

Author details

Antonio Barba* and Juan M. Martinez-Orozco
School of Architecture, Engineering and Design, Universidad Europea de Madrid, Madrid, Spain

*Address all correspondence to: antonio.barba@universidadeuropea.es

References

[1] World Health Organization (WHO). Environmental Noise Guidelines for the European Region. Copenhagen: WHO Regional Office for Europe; 2018

[2] European Environment Agency (EEA). Reported Data on Noise Exposure Covered by Directive 2002/49/EC. Copenhagen: European Environment Agency; 2019

[3] European Topic Centre on Air Pollution, Transport, Noise and Industrial Pollution (ETC/ATNI). Noise exposure scenarios in 2020 and 2030 outlooks for EU 28. In: ETC/ATNI Report 3/2019. Kjeller, Norway: European Topic Centre on Air Pollution, Transport, Noise and Industrial Pollution; 2019

[4] Federal Highway Administration (FHWA). Summary of Noise Barriers Constructed by December 31, 2019. Available from: https://www.fhwa.dot.gov/environment/noise/noise_barriers/inventory [Accessed: 07 January 2022]

[5] Yamamoto K. Japanese experience to reduce road traffic noise by barriers with noise reducing devices. EuroNoise 2015. Maastricht, The Netherlands. 31 May-3 June 2015

[6] Kitamura T, Torii Y. Japanese Experience with Highway Noise and Noise Abatement Measures. Conference on Transportation and Land Use Planning Abroad, Ithaca, New York, July 22-23, 1975 1976.

[7] Maekawa Z. Noise reduction by screens. Applied Acoustics. 1968;**1**:157-173

[8] Kurze UJ, Anderson GS. Sound attenuation by barriers. Applied Acoustics. 1971;**4**:35-53. DOI: 10.1016/0003-682X(71)90024-7

[9] Moreland JB, Musa RS. The performance of acoustic barrier. Noise Control Engineering. 1973;**1**:98-101

[10] Kurze UJ. Noise reduction by barriers. Journal of the Acoustical Society of America. 1974;**55**:504-518. DOI: 10.1121/1.1914528

[11] Pierce AD. Diffraction of sound around corners and over wide barriers. Journal of the Acoustical Society of America. 1974;**55**:941-955

[12] Barry TM, Reagan J. FHWA traffic noise prediction model. In: Report No. FHWA-RD-77-108. Washington, DC: Federal Highway Administration; 1978

[13] Arenas PG. Potential problems with environmental sound barriers when used in mitigating surface transportation noise. Science of the Total Environment. 2008;**405**:173-179. DOI: 10.1016/j.scitotenv.2008.06.049

[14] Sun L, Zhao Y, Zhang J, Chen D, Zhang X. Research and application of noise barriers in highway construction. E3S Web of Conferences. 2021;**233**:01087. DOI: 10.1051/e3sconf/202123301087

[15] European Commission (EC). Noise abatement approaches. In: Future Brief 17. Bristol: Science Communication Unit, UWE; 2017

[16] Conference of European Directors of Roads (CEDR). State of the art in managing road traffic noise: Noise barriers. In: Technical Report 2017-02. Brussels, Belgium: CEDR; 2017

[17] Federal Highway Administration (FHWA). FHWA Highway Noise Barrier Design Handbook. Washington, DC: US Department of Transportation; 2000

[18] Conference of European Directors of Roads (CEDR). Identifying the key characteristics for environmental noise barrier condition measurements. In: Practical Road Equipment Measurement, Understanding and Management (PREMiUM). Brussels, Belgium: CEDR; 2016

[19] Jean P. The effect of structural elasticity on the efficiency of noise barriers. Journal of Sound and Vibration. 2000;**237**:1-21. DOI: 10.1006/jsvi.2000.3059

[20] Wang HB, Cai M, Zhong SQ, Li F. Sound field study of a building near a roadway via the boundary element method. Journal of Low Frequency Noise, Vibration and Active Control. 2018;**37**:519-533. DOI: 10.1177/1461348417725955

[21] He ZC, Li GY, Liu GR, Cheng AG, Li E. Numerical investigation of ES-FEM with various mass redistribution for acoustic problems. Applied Acoustics. 2015;**89**:222-233. DOI: 10.1016/j.apacoust.2014.09.017

[22] Papadakis NM, Stavroulakis GE. Finite element method for the estimation of insertion loss of noise barriers: comparison with various formulae (2D). Urban Science. 2020;**4**:77. DOI: 10.3390/urbansci4040077

[23] Hiraishi M, Tsutahara M, Leung RCK. Numerical simulation of sound generation in a mixing layer by the finite difference lattice Boltzmann method. Computers & Mathematics with Applications. 2010;**59**:2403-2410. DOI: 10.1016/j.camwa.2009.08.073

[24] Nilsson ME, Kaczmarek T, Berglund B. Perceived soundscape evaluation of noise mitigation methods. In: Inter-Noise 2004 Proceedings: 2683-2688. Prague; 2004

[25] Nilsson ME, Andéhn M, Leśna P. Evaluating roadside noise barriers using an annoyance-reduction criterion. Journal of the Acoustical Society of America. 2008;**124**:3561-3567. DOI: 10.1121/1.2997433

[26] Hong JY, Jang HS, Jeon JY. Evaluation of noise barriers for soundscape perception through laboratory experiments. Acoustics-2012. 23-27 April; Nantes, France; 2012

[27] Schröder D, Svensson UP, Vorländer M. Open measurements of edge diffraction from a noise barrier scale model. Proceedings of the International Symposium on Room Acoustics, ISRA 2010. 29-31 August; Melbourne, Australia; 2010.

[28] Li Q, Duhamel D, Luo Y, Yin H. Analysing the acoustic performance of a nearly-enclosed noise barrier using scale model experiments and a 2.5-D BEM approach. Applied Acoustics. 2020;**158**:107079. DOI: 10.1016/j.apacoust.2019.107079

[29] Association Française de Normalisation (NFS) 31089. Code d'essai pour la determination de caracteristiques acoustiques d´ecrans installes en champ libre. La Plaine Saint-Denis, France: AFNOR; 1990

[30] Adrienne Research Team. Test methods for the acoustic performance of road traffic noise reducing devices—Final report. In: European Commission-DGXII-SMT Project MAT1-CT94049. Brussels, Belgium: European Commission; 1998

[31] Clairbois J-P, de Roo F, Garai M, Conter M, Defrance J, Oltean Dumbrava C, et al. Guidebook to Noise Reducing Devices Optimisation. European Project QUIESST (FP7-SST-2008-RTD-1 SCP8-GA2009-233730).

Brussels, Belgium: European Commission; 2012

[32] European Committee for Standardization (EN). EN 1793-4:2015. Road traffic noise reducing devices—Test method for determining the acoustic performance—Part 4: Intrinsic characteristics. In: In Situ Values of Sound Diffraction. Brussels, Belgium: European Committee for Standardization; 2015

[33] European Committee for Standardization (EN). EN 1793-5:2016. Road traffic noise reducing devices—Test method for determining the acoustic performance—Part 5: Intrinsic characteristics. In: In Situ Values of Sound Reflection under Direct Sound Field Conditions. Brussels, Belgium: European Committee for Standardization; 2016

[34] European Committee for Standardization (EN). EN 1793-6:2018+A1:2021. Road traffic noise reducing devices: Test method for determining the acoustic performance—Part 6: Intrinsic characteristics—In situ values of airborne sound insulation under direct sound field conditions. Brussels, Belgium: European Committee for Standardization; 2021

[35] Morcillo MA, Bragado B, Hidalgo A, Cordero R. Proyecto Europeo QUIESST: Diseño y utilización óptima de pantallas acústicas. TecniAcústica 2013. Valladolid, España. 2013. ISBN: 978-84-87985-23-2

[36] Kim C, Cang T, Park Y, Kang M. Test Method for Determining the Acoustic Performance of Noise Reducing Devices Installed on the Top of Highway Noise Barriers. Korea: Korea Expressway Corporation; 2010

[37] International Organization for Standardization (ISO). Acoustics: In Situ Determination of Insertion Loss of Outdoor Noise Barriers of All Types. ISO 10847:1997. Geneva, Switzerland: International Organization for Standardization; 1997

[38] American National Standards Institute (ANSI). Methods for determination of insertion loss of outdoor noise barriers. In: ANSI/ASA S12.8-1998 (R2013). Melville, New York: American National Standards Institute/Acoustical Society of America; 2013

[39] Federal Highway Administration (FHWA). Noise Measurement Handbook. Final Report. Washington, DC: US Department of Transportation; 2018

[40] Anfosso-Lédée F, Steimer V, Demizieux P. In situ methods for the characterisation of noise barriers efficiency. Inter-Noise 2000. 27-30 August; Nice, France; 2000

[41] Parnell J, Samuels S, Tsitsos C. The performance of noise barriers in attenuating road traffic noise. Euronoise 2009. Edinburg, Scotland; 2009

[42] Martinez-Orozco JM, Barba A. Determination of Insertion Loss of noise barriers in Spanish roads. Applied Acoustics. 2022;**186**:108435. DOI: 10.1016/j.apacoust.2021. 108435

[43] Pultznerová A, Šimo J, Grenčík J. Possibilities of evaluating the effectiveness of noise barriers in Slovakia. Applied Sciences. 2021;**11**:10206. DOI: 10.3390/app112110206

[44] Wayson R, MacDonald J, Lindeman W, Berrios M, El-Assar A. Florida Noise Barrier Evaluation and Computer Model Validation. Orlando, Florida: University of Central Florida; 2003

[45] Liu P, Chen S, Wu C. Evaluation on effects of noise barrier defects on their noise reduction efficiencies. Joint

International Conference on Computing and Decision Making in Civil and Building Engineering. June 14-16; Montréal, Canada; 2006

[46] Bragança L, Freitas E, Pinheiro D. Eficacia de barreiras acústicas. TecniAcustica. Gandia, Spain; 2006

[47] Li Q, Duhamel D, Luo Y, Yin H. Improved methods for in-situ measurement railway noise barrier Insertion Loss. Transactions of Nanjing University of Aeronautics and Astronautics. 2018;**35**:58-68. DOI: 10.16356/j.1005-1120.2018.01.058

[48] European Committee for Standardization (CEN). Railway applications. Track. Noise barriers and related devices acting on airborne sound propagation. Test method for determining the acoustic performance. Part 7: Extrinsic characteristics. In situ values of insertion loss. CEN/TS 16272-7:2015. Brussels: European Committee for Standardization; 2015

[49] Bowlby W, Williamson R, Reiter D, Patton C, Pratt G, Kaliski K, et al. Field evaluation of reflected noise from a single noise barrier. In: Research Report 886. NCHRP, TRB. Washington, D.C.: National Academies of Sciences, Engineering and Medicine; 2018. DOI: 10.17226/25297

[50] Palma MJC, Samagaio A. Acoustic performance of a noise barrier coated with an absorptive material. Noise Control Engineering Journal. 2006;**54**:245-250. DOI: 10.3397/1.2219895

[51] Bastian-Monarca NA, Álvarez JP, Reyes CH. Cálculo de pérdida de inserción de barreras acústicas en la Ruta 5 Norte/Sur, tramo concesionado, a partir de mediciones del nivel de presión sonora. INGEACUS 2020 Conference. Valdivia, Chile. 2020

[52] Barba A. Análisis de la eficacia de las pantallas acústicas: Evaluación in situ del comportamiento de las barreras anti-ruido en carreteras. In: Tesis Doctoral. Madrid: Universidad Europea de Madrid; 2017

[53] Cho DS, Kim JH, Choi TM, Kim BH, Manvell D. Highway traffic noise prediction using method fully compliant with ISO 9613: comparison with measurements. Applied Acoustics. 2004;**65**:883-892. DOI: 10.1016/j.apacoust.2004.03.004

[54] Jagniatinskis A, Fiks B, Mickaitis M. Determination of insertion loss of acoustic barriers under specific conditions. Procedia Engineering. 2017;**187**:289-294. DOI: 10.1016/j.proeng.2017.04.377

[55] Watts G. A comparison of noise measures for assessing vehicle noisiness. Journal of Sound and Vibation. 1995;**180**:493-512

Chapter 4

Impact of Noise Pollution during Covid-19: A Case Study of Balasore, Odisha

Bijay Kumar Swain, Chidananda Prasad Das and Shreerup Goswami

Abstract

Activities such as development of industrialisation, urbanisation is a part of our life in the present scenario. During this phase we face a lot of health issues due to noise pollution. Growing of vehicle traffic is one of the major causes towards noise pollution and it affects significantly on the environment. The impact of such pollution had been assessed in 20 major squares (Commercial, residential and silence area) of the Balasore town during and after lockdown imposition of Covid-19. During lockdown period, the noise level of the town was within the permissible limit set by CPCB while before and after lockdown period it was beyond the permissible limit. The demographics and psychophysiological (annoyance, sleeping problem, tiredness, headache, and depression) responses of the participants were collected using standard questionnaires. It was also observed that there were better health conditions among the public (150 participated in the questionnaire) during the lockdown period, then before and after the lockdown phase. It was revealed that socio-demographic factors have no effects on the annoyance level.

Keywords: noise pollution, health issues, Covid-19, equivalent noise level, Balasore

1. Introduction

One of the most common job-related occupational risks is noise and is a global problem. In urban areas it affects the health of people and also the environment. In many reports it has been reported how the people from different part of world are exposed and affected by noise pollution [1–4]. Many studies also reported that there is a corelation between noise and health problems like headache, irritability etc. [5–7]. The main source of noise pollution is vehicular traffic noise or road traffic noise, as reported by many studies [3, 8–12]. Increased noise exposure is known to produce annoyance [5, 13, 14], headaches [15–18], diabetes [19], irritability [20], sleep disturbances [21–26], hypertension [27–30], and problem in blood pressure [31]. Presently, it is a global problem [32].

Again, in many studies, it was also reported about the noise pollution level and its impact on public in world-wide [33–36]. Similarly, in many parts of India, research has been going-on on noise pollution and its impact on human health. In most of the study, it also been reported that the noise levels on Indian road conditions was more than the prescribed noise level set by CPCB [37]. The noise levels of many towns of Odisha are also more than the prescribed limit [38–52]. Silence zones were the most affected by noise pollution, according to Kalawapudi et al. [53], followed by residential, business, and industrial zones. They went on to say that proper city design could help people avoid being exposed to growing noise pollution levels, in Mumbai Metropolitan region. Thakre et al. [54] also discovered a 4.4 and 5.2 dB increase in the morning and evening sessions, respectively, in Nagpur from 2012 to 2019 [54]. The impact noise on bus driver [9], public coming to the park for refreshment [10], Office [55], Bank [56, 57], festivals [11, 41], Industrial areas [58, 59] and workers working in the stone crusher industry [60, 61] has also been reported. Zambon et al. [62] reported about the comparison to the same period in 2019, noise levels in terms of both absolute noise levels (Lden) and hourly noise profiles (median across lockdown period) showed a substantial drop of nearly 6 dB [62], while it was 1–3 dB in Boston metropolitan areas of USA [63] and reduction of 5.1 dB in Ruhr area of Germany [64]. The highest sound levels were found along major roadways, with a logarithmic reduction as distance from the roads increased [63]. Significant outdoor noise fluctuations were discovered, and participants clearly perceived noise variations both in urban and indoor settings, claimed by Caniato et al. [65]. Alias and Alsina-Pages reported that there was a significant reduction in the harmful impact of noise on the population of Milan urban and Rome suburban areas [66].

Now, most of the Indian cities are going to face major threats in the form of noise pollution on public's health. It can affect both physically and mentally on the public's health. But the life changed after the spreading of COVID-19 in whole world. After its existence, first Janata Curfew was coming in to existence followed by the lock-down system. During this period the vehicular traffic noise has been reduced drastically in world-wide. But how much it was reduced is a concern. In this study, an attempt has been made to access the noise levels of the Balasore town before, during and after lockdown phase in different areas. The impact of such noise levels on public's health was also accessed through questionnaire. Suggestive reduction procedures are also given in the present study.

2. Methodology

2.1 Study area

Balasore is one of the famous districts in the state Odisha and situated in the eastern part of the state. It is famous for its cultural heritage, vast sea-beach and many more. It is also famous for Chandipur Sea Beach. The study area is the district head-quarter. As per 2011 census of India, Balasore District has a population of 2,320,529 in 2011 but estimates as per aadhar uidai.gov.in Dec 2020 data as 2,645,403. But the population of the municipality/metropolitan areas was 1,77,751 and city had 1,18,162. The latitude and longitude of the district is 21 29 39 North, 86 55 54 East respectively (**Figure 1**). The monitoring town has elevation of 16 m. the maximum and minimum temperatures are observed to be 31.8 and 21.9 respectively, with an average rainfall of 1706.1 mm, average relative humidity of 71% and speed of 11 km/h. The research area is about 194 km away from the state capital. Different rural roads are connected to this town. Thousands of vehicles along-with number

Figure 1.
Map of India showing the location area of the study area.

of heavy vehicles are flowing on different roads of the town. The town has a very wide commercial areas and lot of people from different regions were depending on this market for their daily needs. The major road of the town also connected with the Chandipur beach, and other religious areas of the district. Thus, heavy rush in vehicle flow has been shown on the town. Every day, thousands of different cars enter and exit the city. The metropolitan environment has a diverse traffic flow. It is one of the busiest municipalities/towns of the state, with a variety of land-use patterns.

Nationwide lockdown (21 days) imposition in India was implemented between 25th March 2020 and 14th April 2020 as Phase 1 and between 15th April and 3rd May 2020 as Phase 2, Phase 3 from 4th May 2020 and 17th May 2020 and last phase (Phase 4) 18th May 2020 to 31st May 2020. Before this nation-wide voluntary public curfew was implemented on 22nd March 2020 for a time period of 14-hour. The same process of lockdown was also implemented in the Balasore town accordingly. Only essential good services are provided to the public. The Unlock phases was came into exist. The first unlock 1.0 came in to exist between 1st June to 30th June 2020. After the month of June 2020, the unlock phases was going on from unlock phase 1 to unlock 21 (1 February 2022 to 28 February 2022). In the present study, the noise levels recorded during unlock phase 1.0 and 2.0, i.e. 1st June 2020 to 30th June 2020 and 1st July 2021 to 31st July 2021. Similarly, the noise level also monitored during December 2019, January

2020 and February 2020 before imposition of the lockdown. During lockdown phase, the noise level had been accessed in the month of May 2020.

2.2 Monitoring sites

At 20 separate locations throughout the town, the acoustic level was measured. All these monitoring stations are divided into three sections such as commercial zone, residential and silence zone. Seven locations from both commercial and residential zones are selected and six stations were selected for silence zone. Some of these locations are belong to the commercial zone, such as Cinema square, Fandi square, Motiganj Bazar, Station square, ITI square, Padhuanpada square, and Policeline square, while others are in silence areas, such as Hospital gate, Durganurshing home, FM college, Zilla school, Near Kendriya Vidyalay (KV), and Police High School and others are in residential areas, such as Mandal bagicha, Near ACPL apartment, Khaparapada New Colony, Rajabagicha, Angargadia, Santikanana and Swastik tower.

2.3 Sampling and data acquisition

The sound level metre Model HD2110L was used to collect acoustic data at each of the 20 sample stations in and around Balasore town. The calibration of the equipment was carried out according to the manufacturer's instructions. The measurements were conducted on working days at street level in and around the chosen locations' major road connections. The instrument was comfortably set in road sides, with the microphone aimed at the source of noise. The equipment was placed 2 m distant from the reflecting object, and the data was gathered while standing 1.5 m above ground level on the roadside. Within 10–20 m gaps, noise levels were measured based on road width. Each station's noise levels were measured in the morning (8–10 a.m.), afternoon (3 p.m.–5 p.m.), and evening (7 p.m.–9 p.m.). The noise levels were measured in four different directions at each station, and one reading was taken every 2 min, for a total of five readings within a 10-minute time frame [67–70]. All of the information is saved on a computer for further study.

For noise level data analysis, noise indices such as Lmin, Lmax, and Leq were calculated. The maximum, minimum, and equivalent noise levels were calculated using all of the recorded data on an excel sheet. The minimal sound pressure level is Lmin, the maximum sound pressure level is Lmax, and the equivalent continuous sound level during that time period is Leq. Again, L10 and L90 refer to sound intensities that are greater than 10% and 90%, respectively.

2.4 Community response

The community reaction was gathered through the use of questionnaires distributed to members of the public going along the various route segments. During the month of March 2020, the public's replies were gathered and recorded on a computer. The questionnaire sent to the participants through whatsapp and in some cases hard copies are also shared and the process was completed during the month of November 2020. One hundred fifty participants have been responded to the questionnaire. This questionnaire was filled out by individuals (those who agreed) who were 18 years old or older. There were two sections to the questionnaire. The first section of the questionnaire is about demographics, while the second section is about various health issues related to the town's acoustic noise. The questionnaire in this study was designed in accordance with Vianna et al. [71], and the questionnaire was constructed appropriately. A total of 150 people from various age groups replied to the questionnaire in this study. The first section contains demographic data such as name, gender, age, educational attainment, and marital status. After minor adjustments by Vianna et al. [71], the second half of the questionnaire was separated into the following sections. The respondents completed the noise sensitivity scale created by Weinstein [72] and Eysenc's personality Inventory (EPI) in this study [73]. Two items given under perception of noise such as aware of noise pollution and environmental noise asked the participants to answer in 5-point Likert scale. The question based on annoyance level and anxiety are also in 5-point Likert scale. Questions are given on hearing condition, sound quality of the environment, personality traits such as aggression, depression, stability, working ability, tiredness and drowsy, sensitivity, relaxation, developing symptoms, and on health risk. This part asks about how people perceive noise from things like road traffic and other sources, and the answer is either Yes or No. High, Moderate, and Little annoyance in regard to noise sources; Noise exposure effects (hearing loss, sleep disturbances, headaches, fatigue, drowsiness, and other illnesses); Hearing condition (Excellent, Good, Moderate, and Poor); environmental sound quality (Normal, Moderate, and Noisy); and environmental perception (Yes, No, and Undecided) [9, 71, 73]. The Chi-square test in SPSS 20.0 was used to look into the correlations between demographic characteristics and annoyance, and other environmental factors and the ANNOVA test was used to look into the association between noise exposure and the probable impacts of that noise on this community. At a significance threshold of 0.05, the relationship between individual and combination socio-demographic characteristics was examined. The datasets were analysed using SPSS software (20.0).

3. Results and discussion

3.1 Studies related to zone specific noise

The average noise levels of the 20 stations of different categories have been accessed and presented in **Table 1** (Before Lockdown), **Table 2** (during lockdown) and **Table 3** (Unlock phases). The data collected during the month December 2019 and January and February 2020 are considered as pre-lockdown phase. 17th March 2020 to 31st May 2020 considered as lockdown period. After 1st June 2020 it is considered as unlock phases or after lockdown period. The comparative monthly variation of equivalent noise levels of these areas having different land use type is presented in **Figure 2**. The figure clearly depicts that there is a sharp trend of noise levels of the town during three phases of the lockdown. It also demonstrates that the noise level during the lock down phases is very low than unlock and before lockdown phases. The monthly noise variation of all the stations is depicted in the **Figures 2–10**. In each figure, first three belongs to the monthly noise level before lockdown period, while the fourth one belongs to the lockdown period and the last portions belong to the unlock phases.

The **Table 1** clearly depicts that the noise levels for commercial zone ranged from 57.7 to 91.6 dB, 57.3 to 91.6 dB and 56.8 to 93.7 dB for morning, afternoon and evening hour respectively. Similarly, the noise level for silence zone ranged from 57.4 to 90.3 dB; 58.8 to 91.7 dB and 53.5 to 91.6 dB during the morning, afternoon and evening hour respectively and from 55.9 to 85.6 dB; 56.8 to 89.8 dB and 50.2 to 91.6 dB

Sl. No.	Name of the square	Morning (8–10 am)						Afternoon (3–5 pm)						Evening (7–9 pm)					
		Max	Min	L10	L50	L90	Leq	Max	Min	L10	L50	L90	Leq	Max	Min	L10	L50	L90	Leq
1	Cinema sq	88.9	61.6	82.6	75.8	66.4	80.5	90.4	59.8	82.7	75.6	65.2	81.1	89.3	58.4	83.6	76.9	67.5	81.5
2	Fandi sq	89.2	62.4	81.4	76.2	67.3	79.8	91.6	60.7	82.1	75.2	64.8	80.5	90.8	59.5	83.7	75.4	68.2	79.7
3	Motiganj Bazar	87.7	57.7	80.8	75.1	64.9	79.6	88.7	59.3	81.8	76.3	64.4	81.7	90.3	59.8	82.5	75.3	67.4	79.4
4	Station sq	86.6	59.9	80.3	75.4	63.7	80.3	91.4	60.2	80.7	74.5	62.7	80.3	90.7	60.8	81.4	76.7	70.3	78.9
5	ITI sq	86.8	61.6	80.3	74.6	65.8	78.4	88.6	58.4	79.9	72.7	63.6	77.4	89.3	56.8	78.7	72.6	65.5	75.7
6	Padhuanpada sq	91.6	58.8	79.7	72.1	62.7	77.3	90.8	57.3	80.8	74.6	65.8	78.6	93.7	57.5	81.4	76.7	68.4	79.7
7	Policeline sq	86.4	58.5	79.5	71.6	61.8	77.2	87.9	57.4	79.7	72.2	65.6	75.7	88.2	58.6	79.7	72.4	66.6	75.5
8	Hospital gate	87.4	62.7	79.3	72.8	65.4	76.3	88.3	61.3	80.6	73.7	64.8	78.2	91.6	59.7	80.3	74.5	65.5	78.4
9	Durga nursing home	86.7	61.8	77.5	70.4	63.7	73.8	88.1	60.9	79.1	72.2	65.7	75.4	90.8	58.4	79.1	72.5	65.8	75.6
10	FM college	89.8	64.3	79.4	73.6	67.8	76.0	91.7	62.2	80.3	75.8	70.1	77.7	90.3	56.6	80.8	76.3	71.7	77.7
11	Zilla School	86.2	57.7	77.3	71.4	62.4	75.4	89.9	59.6	80.4	75.2	68.4	77.7	86.5	53.5	77.3	72.8	64.9	75.5
12	Near KV	90.3	57.4	75.5	70.1	62.9	72.9	91.6	58.8	78.3	72.8	64.4	76.3	85.9	54.8	77.7	72.4	65.8	74.9
13	Police HS	86.6	59.5	76.2	69.3	62.6	72.6	87.9	60.3	77.3	70.6	63.5	74.0	84.3	53.8	77.8	71.4	67.2	73.4
14	Mandal bagicha	82.8	55.9	74.2	67.2	59.1	71.3	84.4	56.8	75.7	68.9	60.4	73.1	85.9	55.2	72.2	66.4	59.2	69.4
15	Near ACPL Apartment	84.7	56.6	73.2	66.2	59.4	69.6	86.2	57.9	72.8	65.4	60.3	68.2	88.4	54.8	71.4	65.3	57.7	68.6
16	Khaparapada New Colony	82.8	56.3	73.7	65.9	59.4	69.5	85.8	58.6	74.4	67.7	60.9	70.9	85.2	50.2	71.6	65.1	58.4	68.2
17	Rajabagicha	85.6	64.7	76.8	71.3	66.6	73.2	87.3	61.5	77.4	71.9	64.8	74.7	91.6	60.6	75.9	70.3	64.7	72.5
18	Angargadia	85.1	61.3	75.1	69.8	64.8	71.7	89.8	59.4	75.6	70.2	63.9	72.6	90.4	54.2	73.2	69.1	62.7	71.1
19	Santikanan	83.8	57.6	74.7	69.3	64.9	71.0	86.1	58.3	74.7	69.7	62.8	72.2	85.7	50.6	70.1	64.2	59.3	66.3
20	Swastik tower	84.1	62.6	73.9	68.4	64.7	69.9	87.3	60.9	74.5	69.8	64.1	71.7	89.2	50.3	70.3	63.4	58.6	65.8

Table 1.
Noise levels in dB at different traffic squares of Balasore town during different time interval (pre-lock down phase).

Sl. No.	Name of the square	Morning (8–10 am)						Afternoon (3–5 pm)						Evening (7–9 pm)					
		Max	Min	L10	L50	L90	Leq	Max	Min	L10	L50	L90	Leq	Max	Min	L10	L50	L90	Leq
1	Cinema sq	62.4	43.5	57.7	53.7	50.5	54.6	61.9	42.1	57.3	53.4	49.9	54.4	60.7	40.3	54.5	50.5	48.4	51.2
2	Fandi sq	67.8	42.6	56.1	52.8	48.6	53.8	63.5	41.7	56.5	52.7	48.8	53.7	64.7	41.1	55.1	51.6	48.2	52.5
3	Motiganj Bazar	70.1	40.1	58.4	54.4	48.4	56.2	69.5	40.2	56.1	52.5	48.6	53.5	66.8	43.8	55.7	52.8	49.9	53.4
4	Station sq	66.3	41.4	56.9	53.8	49.7	54.7	70.2	40.6	55.9	52.4	48.4	53.4	64.6	42.7	54.8	51.7	49.4	52.2
5	ITI sq	64.9	40.3	56.1	51.8	47.4	53.2	78.5	40.4	55.9	51.4	46.8	52.8	67.6	41.7	54.3	50.7	47.8	51.5
6	Padhuanpada sq	71.7	41.8	56.4	52.2	48.2	53.4	81.4	40.3	55.7	51.8	47.4	53.0	65.9	41.9	54.7	50.2	47.6	51.1
7	Policeline sq	64.4	40.1	55.5	52.1	48.3	53.0	65.4	40.7	54.6	51.8	46.9	52.8	63.9	40.9	53.8	50.7	46.5	51.6
8	Hospital gate	72.8	53.8	61.4	57.7	55.7	58.3	70.7	48.7	60.2	55.7	52.9	56.7	78.3	51.6	63.8	58.6	52.8	60.7
9	Durga nursing home	67.3	44.8	60.3	56.3	52.5	57.4	65.8	45.1	59.4	54.7	50.1	56.2	68.7	47.6	55.7	53.6	49.4	54.3
10	FM college	66.1	44.2	54.4	52.4	48.8	52.9	60.8	42.8	56.1	52.6	50.6	53.1	62.8	42.7	53.6	49.5	44.7	50.9
11	Zilla School	65.3	40.8	54.8	51.2	47.6	52.1	61.2	44.6	54.7	50.6	47.8	51.5	66.5	44.8	53.3	49.2	45.1	50.4
12	Kendriya vidyalaya	62.8	41.7	53.6	50.4	47.4	51.1	61.7	43.8	54.2	49.6	46.8	50.6	63.9	42.5	53.1	49.1	45.6	50.1
13	Police HS	64.9	42.8	53.7	50.4	47.8	51.0	60.8	42.8	53.4	48.6	45.8	49.6	67.4	41.6	52.9	48.7	44.3	50.0
14	Mandal bagicha	58.6	40.2	51.7	45.4	43.7	46.5	57.6	39.5	49.5	44.3	42.6	45.2	58.1	38.3	49.2	44.1	42.2	44.9
15	Near ACPL Apartment	60.7	41.4	50.3	45.2	43.6	46.0	56.7	40.2	48.8	44.2	42.3	44.9	57.5	39.4	48.2	44.4	41.8	45.1
16	Khaparapada New Colony	71.9	40.7	49.7	45.6	42.8	46.5	63.3	41.1	48.5	44.5	42.8	45.1	56.8	40.2	47.6	43.7	41.4	44.4
17	Rajabagicha	59.5	40.3	50.2	45.7	43.8	46.4	58.4	40.7	48.1	44.1	43.7	44.5	57.8	40.5	47.5	43.5	41.8	44.1
18	Angargadia	67.3	42.3	51.1	47.7	44.7	48.4	59.7	40.2	49.8	44.6	43.3	45.4	57.2	38.4	46.5	43.7	42.7	43.9
19	Santikanan	56.6	40.7	50.3	45.6	43.3	46.5	58.3	39.4	48.6	43.8	40.8	44.8	56.1	39.7	46.7	43.3	42.4	43.6
20	Swastik tower	56.5	40.2	49.5	45.1	42.9	45.9	57.8	38.4	47.3	43.1	49.5	43.2	56.9	38.6	46.2	43.1	41.7	43.4

Table 2.
Noise levels in dB at different traffic squares of Balasore town during different time interval (during lock down phase).

Sl. No.	Name of the square	Morning (8–10 am)						Afternoon (3–5 pm)						Evening (7–9 pm)					
		Max	Min	L10	L50	L90	Leq	Max	Min	L10	L50	L90	Leq	Max	Min	L10	L50	L90	Leq
1	Cinema sq	91.3	55.6	73.8	69.6	64.8	71.0	90.7	56.7	74.7	70.1	65.8	71.5	90.3	54.8	75.3	71.7	66.6	73.1
2	Fandi sq	92.5	54.2	73.3	70	65.2	71.2	91.6	60.7	74.1	70.1	65.3	71.5	90.1	54.2	74.6	70.8	65.3	72.3
3	Motiganj Bazar	90.3	52.8	73.1	68.5	63.7	70.1	88.7	59.3	73.6	69.6	64.4	71.1	88.4	52.9	74.7	70.4	64.2	72.4
4	Station sq	88.5	54.8	72.5	68.7	64.3	69.9	91.4	60.2	73.2	69.4	63.7	71.0	85.2	51.7	73.2	69.8	63.3	71.5
5	ITI sq	87.1	54.7	72.2	68.4	64.7	69.4	88.6	58.4	72.5	68.7	62.9	70.3	84.6	52.5	73.7	68.7	63.6	70.5
6	Padhuanpada sq	90.5	53.6	73.7	69.2	63.5	71.1	90.8	57.3	72.7	69.5	63.1	71.1	83.9	50.5	73.1	69.2	63.2	70.9
7	Policeline sq	89.5	55.5	71.6	67.7	62.4	69.2	87.9	57.4	72.4	69.1	63.8	70.4	83.7	51.6	72.7	68.5	61.8	70.6
8	Hospital gate	88.7	56.8	75.7	70.7	64.4	72.9	88.3	61.3	75.2	71.1	65.8	72.7	86.4	55.2	72.5	68.3	64.7	69.4
9	Durga nursing home	85.3	56.2	70.2	66.2	62.5	67.3	88.1	60.9	72.7	67.3	64.6	68.5	81.8	52.8	72.1	67.4	63.3	68.8
10	FM college	90.4	56.3	70.4	65.4	62.3	66.6	91.7	62.2	72.3	67.2	64.4	68.3	83.7	53.6	72.4	68.1	63.5	69.5
11	Zilla School	85.3	51.9	70.3	65.2	62.6	66.3	89.9	59.6	72.1	66.6	62.8	68.1	82.9	52.5	72.5	67.6	62.4	69.4
12	Near KV	81.1	54.7	70.3	65.4	62.2	66.6	91.6	58.8	71.8	66.2	63.3	67.5	84.6	52.6	71.1	66.5	61.4	68.2
13	Police HS	82.7	54.8	69.5	64.3	60.5	65.7	87.9	60.3	71	65.9	63.5	66.9	83.5	52.1	71.4	66.2	61.3	68.0
14	Mandal bagicha	77.7	52.4	65.3	60.2	56.5	61.6	75.8	48.5	62.4	58.7	54.5	59.8	72.7	51.8	61.6	57.4	53.6	58.5
15	Near ACPL Apartment	78.9	51.9	65.1	59.8	57.8	60.7	72.6	47.7	62.4	57.3	53.3	58.8	70.3	49.3	61.4	57.2	53.3	58.4
16	Khaparapada New Colony	80.5	52.8	64.7	59.5	57.3	60.5	70.7	50.3	62.1	57.4	53.6	58.7	71.9	45.7	61.7	56.5	52.4	58.0
17	Rajabagicha	85.6	60.1	65.5	60.4	56.4	61.9	71.9	49.6	63.6	60.4	55.8	61.5	75.7	49.5	60.8	56.2	52.2	57.5
18	Angargadia	85.9	53.7	64.6	60.1	56.2	61.4	70.8	48.8	62.1	58.3	54.2	59.4	71.1	48.6	60.4	56.3	52.4	57.4
19	Santikanan	84.7	54.8	64.2	60.1	56.5	61.2	73.5	47.5	61.5	56.7	52.6	58.1	70.8	50.1	60.7	56.1	52.7	57.2
20	Swastik tower	82.9	55.6	63.5	59.7	55.6	60.8	71.4	46.2	61.2	56.2	53.2	57.3	70.8	48.4	60.4	55.5	51.8	56.8

Table 3.
Noise levels in dB at different traffic squares of Balasore town during different time interval (post-lock down phase).

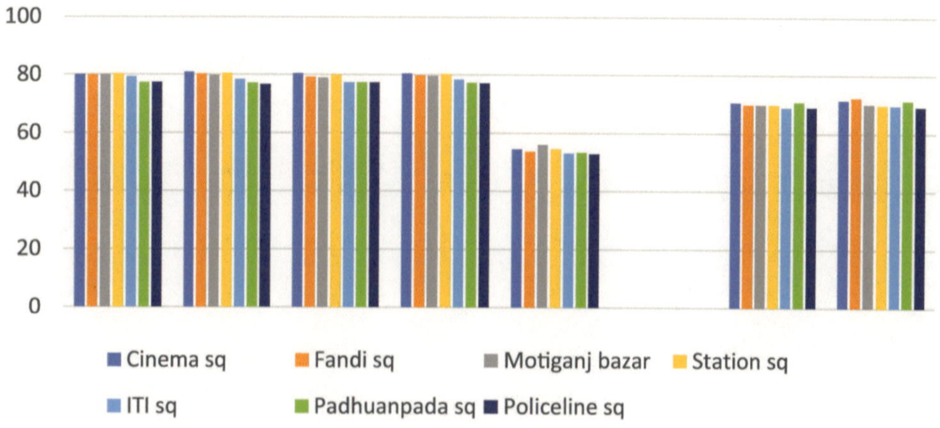

Figure 2.
Comparative equivalent noise level (first three are before lock-down, middle on during lock-down and the last twos are during unlock phase) for the morning hour of commercial zone.

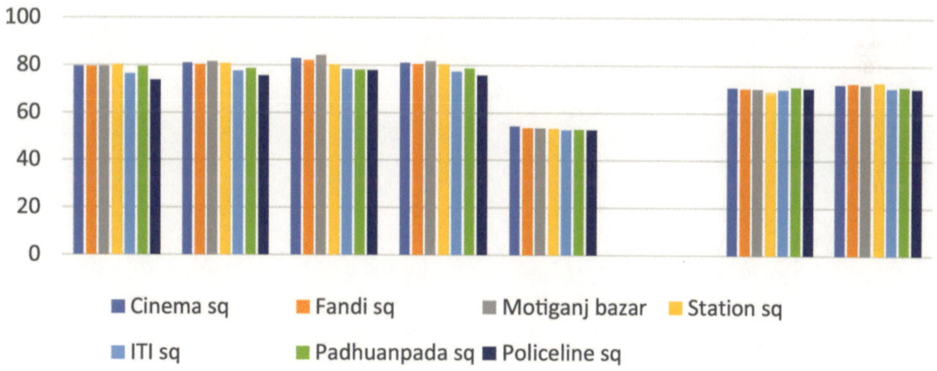

Figure 3.
Comparative equivalent noise level (first three are before lock-down, middle on during lock-down and the last twos are during unlock phase) for the afternoon hour of commercial zone.

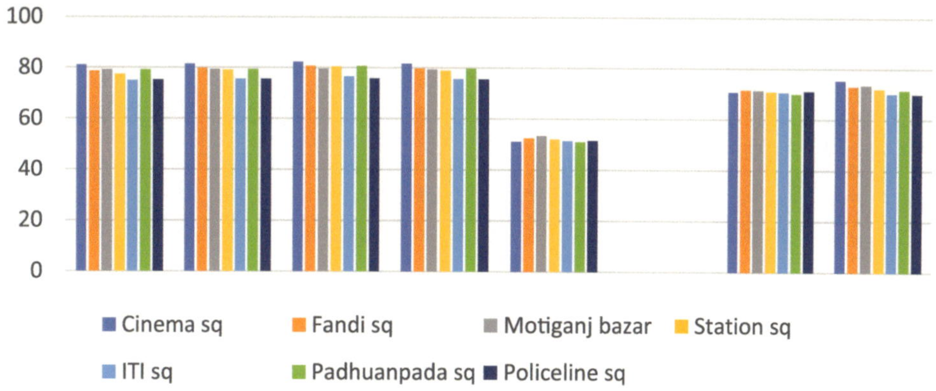

Figure 4.
Comparative equivalent noise level (first three are before lock-down, middle on during lock-down and the last twos are during unlock phase) for the evening hour of commercial zone.

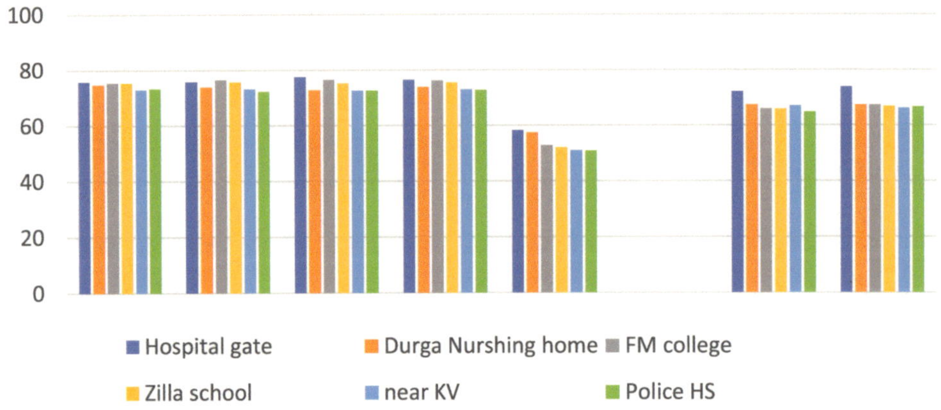

Figure 5.
Comparative equivalent noise level (first three are before lock-down, middle on during lock-down and the last twos are during unlock phase) for the morning hour of silence zone.

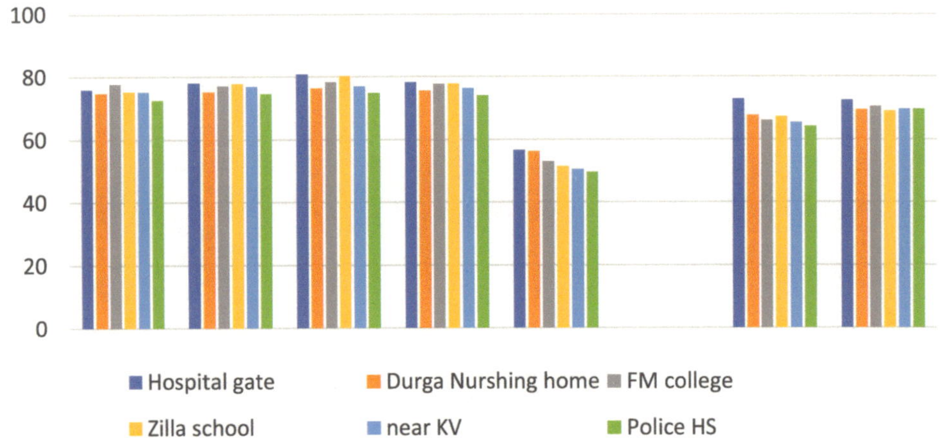

Figure 6.
Comparative equivalent noise level (first three are before lock-down, middle on during lock-down and the last twos are during unlock phase) for the afternoon hour of silence zone.

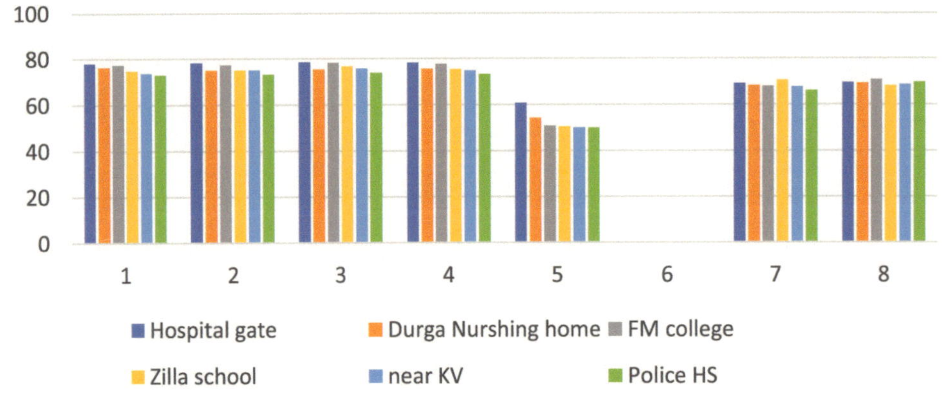

Figure 7.
Comparative equivalent noise level (first three are before lock-down, middle on during lock-down and the last twos are during unlock phase) for the evening hour of silence zone.

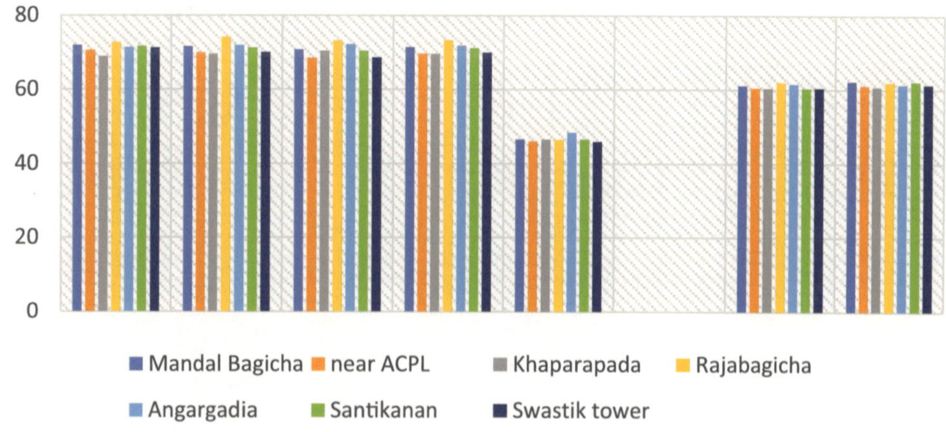

Figure 8.
Comparative equivalent noise level (first three are before lock-down, middle on during lock-down and the last twos are during unlock phase) for the morning hour of residential zone.

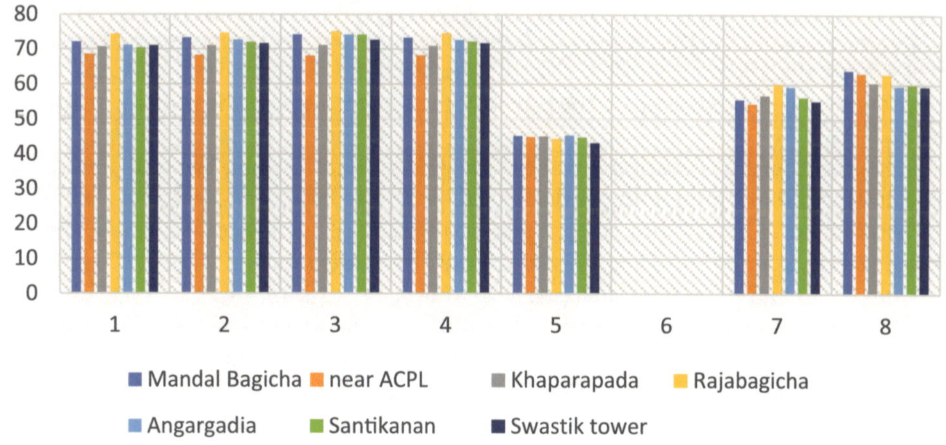

Figure 9.
Comparative equivalent noise level (first three are before lock-down, middle on during lock-down and the last twos are during unlock phase) for the afternoon hour of residential zone.

during the morning, afternoon and evening hour respectively for residential zone. It can be summarised that the noise for all zones before lockdown period had a ranged from 55.9 to 91.6 dB; 56.8 to 91.7 dB and 50.2 to 93.7 dB during the morning, afternoon and evening hour respectively. **Table 1** also clearly depicted that for all time the noise level ranged from 50.2 to 91.7 dB during before lock-down phase (**Table 1**).

Similarly, **Table 2** clearly demonstrated that all zones lie in the range of 38.3 to 81.4 dB during lockdown period (**Table 2**) and then the range gradually increased to 45.7 to 92.5 dB during unlock period (**Table 3**). The equivalent noise levels of all zones lie in the range of 65.8 to 81.7 dB (**Table 1**); reduced to 43.2 to 60.7 dB during lock-down period (**Table 2**) and the range then gradually increased from 56.8 to 73.1 dB during unlock period (**Table 3**). The permitted limit for the said locations, as defined by the CPCB for Indian road conditions, is 65 dB during the day and 55 dB at night [37]. During the day time, the noise level exceeded the permitted limit [74–78]. The noise level during unlock phase and before imposing lockdown was beyond the

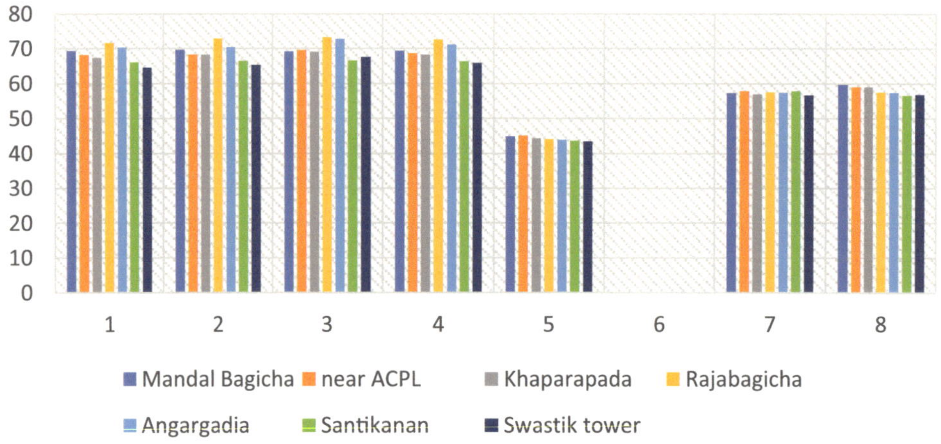

Figure 10.
Comparative equivalent noise level (first three are before lock-down, middle on during lock-down and the last twos are during unlock phase) for the evening hour of residential zone.

permissible limit in the present study. It was reported that, if the exposure of noise level is more than 80 dB (A), then risk of hypertension will increase [34]. More research is needed to investigate the effect of such noise level on the public's health in future study.

From the monthly variation it was demonstrated that the noise levels of residential areas during the morning hour are decreased by a noise level of 24.8 dB and then it increased up to 15.1 dB during unlock phase in Mandal Bagicha area. Similarly, the noise levels in other monitoring areas decreased by a noise level of 23.5, 23, 26.8, 23.3, 24.5 and 24 dB and then it was increased up to 14.7, 14, 15.5, 13, 14.7 and 19.95 dB for ACPL, Khaparapada, Rajabagicha, Angargadia, Santikanan and Swastik tower, respectively (**Table 1**). During afternoon hour, the maximum reduction of noise level was noticed at Rajabagicha area (30.2 dB) and then it increased up to 17 dB during the unlock phase. From the **Table 1**, it clearly depicts that maximum noise reduction between before lock-down phase and during unlock phase was observed at commercial zone (more than 25.5 dB) followed by residential zone (more than 25 dB) and silence zone (more than 22 dB). All these data are mentioned here are in an average data. Similarly, the maximum growing noise level between during lock-down and unlock phase was also noticed at commercial zone (more than 17.8 dB) and followed by silence zone (16.7 dB) and residential zone (14.1 dB). During evening hour and at Padhuanpada square maximum noise reduction i.e., 28.6 dB was noticed, while the lowest reduction was at 16.3 dB during morning hour at Durga Nursing home. Again, maximum increase of noise level was noticed at Cinema square (21.9 dB) during the evening hour, while the minimum increase noise was noticed at Hospital gate (8.7 dB) during the evening hour also.

Table 1 also clearly depicted that the equivalent noise level during before lock down phase ranged from 77.2 to 80.5 dB; 75.7 dB to 81.7 dB and 75.5 to 81.5 dB for morning, afternoon and evening hour respectively. But the noise level during the lock-down phase ranged from 53.9 to 56.2 dB; 52.8 to 54.4 dB and 51.1 to 53.4 dB during the morning, afternoon and evening hour respectively (**Table 2**). Similarly, the noise level during the unlock phase ranged from 69.2 to 71.2 dB; 70.3 to 71.5 dB and 70.5 to 73.1 dB during the morning, afternoon and evening hour respectively (**Table 3**). The noise level at silence zone ranged from 72.6 dB to 76.3 dB; 74 to

78.2 dB; 73.4 to 78.4 dB during before lock down phase; 51 to 58.3 dB; 49.6 to 56.7 dB and 50 to 60.7 dB during lock-down phases at morning, afternoon and evening hour and 65.7 to 72.9 dB; 66.9 to 72.7 dB and 68 to 69.5 dB during morning, afternoon and evening hour respectively. In the residential areas it ranged from 69.5 to 73.2 dB; 68.2 to 74.7 dB and 65.8 to 72.5 dB during before lock down phase; in lock-down phase the noise level ranged from 45.9 to 48.4 dB; 43.2 to 45.4 dB and 43.4 to 45.1 dB and in unlock phase it ranged from 60.5 to 61.9 dB; 57.3 to 61.5 dB and 56.8 to 58.5 dB in the morning, afternoon and evening hour respectively.

In location wise, **Tables 1–3** clearly depicts the noise variation in all the monitoring stations. These Tables demonstrated that there is a reduction of 25.9 dB, 26.7 dB; 30.3 dB in three different monitoring hours for site 1 of commercial zone. Conversely, the reduction is almost 26, 26.8 and 27.2 dB for site 2, 23.4, 28.2, 26 dB for Site-3 and a similar trend was found in all other monitoring sites belong to commercial zone. In the commercial zone the minimum noise reduction ranged from 22.9 to 28.2 dB and 23.9 to 30.3 dB during afternoon and evening hour of the commercial zone. Again, the reduction of noise level ranged from 16.4 to 23.3 dB; 19.2 to 26.2 dB and 17.7 to 26.8 dB of silence zone and from 23 to 26.8 dB; 23.3 to 30.2 dB and 22.4 to 28.4 dB of residential zone during morning, afternoon and evening hour respectively. This result clearly depicted that there is almost same trend in the noise level reduction, both in commercial and residential zone of the town. The minimum noise level reduction was more than 15 dB and found at silence zone of the town and clearly depicted that due to the nationwide lock-down imposition, there was a sharp reduction in the noise level. It will impact the environment in a positive manner.

In comparison between Leq values of a particular sites during the lock-down period with unlock phase, there was sharp increase in the noise levels of each location. Noise levels from 13.9 to 17.7 dB was increased during the morning hour in between lockdown and unlock phases. Similarly, the noise levels increased by 17.1 to 18.1 dB and 19 to 21.9 dB in afternoon and evening hour of commercial zone. Again, the increased noise level ranged from 9.9 to 15.5 dB, 12.3 to 17.3 dB and 8.7 to 19 dB of silence zone and ranged from 13 to 15.5 dB, 13.3 to 17 dB and 13.3 to 13.6 dB of the residential zone during morning, afternoon and evening hour, respectively. Due to slight relaxation provided by the local administration, there was a sharp increase in noise level of the town. This trend was more commercial zone. In the present study it is also reported that there is no relation between the different monitoring hours and the situation i.e., before imposing lockdown and after imposing and lifting the lockdown phase of the town. But there is a good association between different areas such as residential, commercial and silence zone with unlock and before lock down phase of the town (**Table 4**). In case of monthly noise level variation with different phases of the lockdown situation there is also good association between them and is presented in the **Table 5**.

In the present study, it was found that the noise level in the residential areas is growing on due to imposition of lockdown in the town. Due to lockdown, the commercial areas of the town and for such the people are selling different grocery items in the different parts of the residential areas. The open shops are instantly made on the roadside and there is slight gathering around such place. These shops are opened from 7 am to 7 pm during the unlock phase while it was opened from 7 am to 2 pm during the lockdown phase. Around the market or shop area there was gathering and due to which, the noise level during the unlock phase was raising. Again, during the unlock phase, the noise level suddenly increased due to immediate rush in different parts of the town, due to purchase of goods for their house. They creating a such situation unnecessarily by gathering around the temporary shops.

Source	Type III sum of squares	df	Mean square	F	Sig.
Corrected model	34.837[a]	2	17.419	138.614	0.001
Intercept	13.532	1	13.532	107.684	0.001
Unlock	3.144	1	3.144	25.020	0.001
BeforeLockdown	1.186	1	1.186	9.441	0.003
Error	7.163	57	0.126		
Total	282.000	60			
Corrected Total	42.000	59			

[a]R squared = 0.829 (Adjusted R squared = 0.823).

Table 4.
Two way ANNOVA analysis for equivalent noise levels during unlock and before lockdown phases with different areas.

		Sum of squares	df	Mean square	F	Sig.
December	Between groups	738.064	2	369.032	92.802	.001
	Within groups	226.663	57	3.977		
	Total	964.727	59			
January	Between groups	755.856	2	377.928	87.198	.001
	Within groups	247.046	57	4.334		
	Total	1002.902	59			
February	Between groups	780.520	2	390.260	68.810	.001
	Within groups	323.277	57	5.672		
	Total	1103.797	59			
May	Between groups	868.172	2	434.086	98.290	.001
	Within groups	251.732	57	4.416		
	Total	1119.904	59			
June	Between groups	1655.407	2	827.704	232.770	.001
	Within groups	202.686	57	3.556		
	Total	1858.093	59			
July	Between groups	1464.398	2	732.199	209.703	.001
	Within groups	199.022	57	3.492		
	Total	1663.419	59			

Table 5.
One way ANNOVA analysis for monthly equivalent noise levels with different areas.

In case of silence zone, schools and hospitals were taken in the present study. All the monitoring stations were located along the main road. College and school squares are also along the road of different hospitals. Many private clinics and hospitals are also very close to the schools and colleges of the town. During lockdown, many shops, schools and colleges and other establishments were closed. All medicinal shops are opened throughout the day time. But vehicles are flowing on the road due to health matter. Continuous flowing of many vehicles including heavy vehicles on the road are controlled, but running of the two wheelers, ambulances and responsible for the noise levels of the town.

3.2 Community responses

During the month of March 2020, the public's replies were gathered and recorded on a computer. The questionnaire was supplied to the participants both in online and offline mode. Those are expertise in the android mobile phones or in their PC or laptop they are responded to the questionnaire through online mode. Those are not comfortable in using these devices, asked the researcher to provide the such through offline mode and also provided to them as such. After getting their responses, it was then transferred in to MS Excel for its further analysis.

The questionnaire was completed by 150 people, as mentioned in the content and methods section. The average age of the responders was 37.8 years old, with a standard deviation of 9.4 years. **Table 6** lists the various personal characteristics of the individuals. **Table 6** clearly depicts that the majority of the participants are male respondent (59.3%), with 68.7% of the total completing their education at the graduation level. In the present study, majority of the participants are employed. Majority of the participants (78.7%) participated in this survey work are married. In the present study most of the young generation (48%) between the age of 18 to 30, responded to this questionnaire.

The Pearson Chi-square of noise discomfort to different demographic characters is shown in **Table 7**. **Table 7** clearly depicts that there is a good association between annoyance and gender of the present work. Again, there is no direct relationship between annoyance and other demographic characteristics, according to the data.

In this study it was found that 36.7% individuals were extremely irritated, while 39.5% remain silent. In a study conducted by Alimohammadi et al. [73] on White-collar

Sl. No.	Characteristics	Variables	Response in %
1	Gender	Male	59.3%
		Female	40.7%
2	Age	18–30	48%
		31–45	36.7%
		More than 45	15.3%
3	Marital status	Unmarried	21.3%
		Married	78.7%
4	Qualification	Under matric	7.3%
		Matric	5.3%
		Intermediate	18.6%
		Graduation	68.7%
5	Occupation	Employed	51.3%
		Self-Employed	28%
		Un-employed	20.7%

Table 6.
Personal characteristics of respondents (in percentage).

Demographic characters	X^2	df	Asymp. Sig. (2-sided)
Gender	15.622	4	0.004
Age	7.384	8	0.496
Marital status	6.968	4	0.138
Qualification	11.698	12	0.470
Occupation	8.679	8	0.370

Table 7.
Relation between demographic character and annoyance.

employees in Teharan, it was discovered that married people were more irritated than unmarried people. But in the present study it was contradicted that result (p = 0.217).

The participants' perceptions on noise, health issues, hearing conditions, sound quality of the environment, environmental problems, opinion of participants on noise preventability, sensitivity to noise, annoyance, and the importance of controlling the town's noise were all examined in the current study and presented in the **Table 8**.

On awareness towards road traffic noise pollution, majority of the participants (59.3%) were aware of it. More than 30% respondents were strongly aware about the noise pollution, which is also a good sign for the society. Regarding health issues majority respondents (36.7%) opined about a little impact of noise pollution on their health, while 24.6% respondents remain silent and only 18% viewed that they suffered moderately by the noise pollution. On hearing condition most of the participants (38%) were in moderate condition, while 30% responded as good in condition. Only 27.3% opined that their hearing condition was not so good or in bad condition. How much the hearing problem is affected is not studied in the present study. The researcher aimed to conduct the audiometry study of these respondents very soon to know their actual level of hearing in the next study. Noise induced hearing loss is also the most frequently recognised occupational disease in many countries [79–81].

The sound quality of the town was not so good as per the response of the participants. Due to such issues, they face a lot of problems (56.7%) in their day-to-day life. According to the findings, 40.7% of the participants suffered illness, while most of them faced headache (54.7%) due to road traffic noise. How much it affects the public health and what are the possible symptoms are developed is to be investigated in the next phase of study. Majority of the respondents (57.3%) responded that they annoyed often. Running of vehicles (89.3%) is the major source of pollution, followed by railway (76.7%), two wheelers (75%), honking (70.7%) (**Table 8**).

The acoustic quality of the area was described as noisy by the majority of the participants (60%). According to the study, majority of the of interviewees felt that road traffic noise was polluting the environment. When the participants' knowledge was assessed, most of them said that road traffic noise poses a significant health risk. Noise pollution upset 67.3% of the participants, while 58.7% were sensitive to noise and 60% found it difficult to relax in these situations. More than 48% felt depressed, 82% were felt tired, 48.7% were not working in a stability manner. It may be due to the effect of the noise pollution.

The chi-square test was used to determine the relationship between age and annoyance in this study, and no link was found at p = 0.01. However, there is a strong association between annoyance and gender (p = 0.004). There was also a link between work place noise levels and annoyance, according to Allomohammadi et al. [73].

Personality traits		Number	Percentage
Perception of noise (aware of noise pollution)	Strongly agree	46	30.7
	Agree	89	59.3
	Neutral	6	4
	Disagree	9	6
	Strongly disagree	—	—
Health issues	High	10	6.7
	Moderate	27	18
	A little	55	36.7
	No feeling	21	14
	Neutral	37	24.6
Hearing condition	Excellent	1	0.7
	Good	51	34
	Moderate	57	38
	Bad	41	27.3
	Worst	—	—
Sound quality of the environment	Normal	6	4
	Moderate	54	36
	Noisy	90	60
Environmental problems	Strongly agree	29	19.3
	Agree	85	56.7
	Neutral	10	6.7
	Disagree	26	17.3
	Strongly disagree	—	—
Developing symptoms	Feeling ill	61	40.7
	Headache	82	54.7
	Respiratory problems	7	4.7
	Eye irritation	—	—
Annoyance	Never	14	9.3
	Occasionally	5	3.3
	Sometimes	10	6.7
	Often	86	57.3
	Always	35	23.3
Source of noise pollution	Mobile phones	35	23.3
	Running of vehicles	134	89.3
	Honking	106	70.7
	Railway	115	76.7
	Two wheelers	117	75

Table 8.
Participant's perception towards different aspects of noise pollution.

But this result is similar to the present study. It can be said that occupation is not a good characteristic towards annoyance. According to reports, there is no correlation between age, education, or marital status and the town's level of annoyance. The current study's findings are comparable to those of Ohrstrom et al. [82], who found that age, sex, and other characteristics do not explain differences in annoyance between people and is very similar to the results of the present study. However, it has been reported in many research that annoyance is the most vulnerable consequence of traffic noise exposure [83, 84], which contradicts the findings of the current study in many circumstances.

There is good association between gender and drowsiness of the public ($p = 0.015$) (**Table 9**). **Table 9** also demonstrates that there is an association between drowsy and qualification. **Table 10** depicts that there is an association between relaxation and gender ($p = 0.001$) and age ($p = 0.006$). Most of the demographic characters have a good association with noise sensitivity (**Table 11**). Noise sensitivity has a good association with gender ($p = 0.001$), age ($p = 0.005$), marital status ($p = 0.001$) and qualification ($p = 0.038$) of the present study (**Table 11**). **Table 12** reveals that both gender ($p = 0.001$) and marital status ($p = 0.001$) has an association with anxiety of the noise pollution (**Table 12**). Gender is not a significant element in the influence of noise concern, according to certain studies [5, 85]. Similar results also depicted in the present study. There was also a link between the individuals' age and sleep problems ($p = 0.046$). It was also said that age is not a significant factor when it comes to the effects of noise exposure [5, 80]. Increased parent-reported sleep issues were identified in the few studies that looked at the link between noise and child/adolescent sleep [23, 82]. Sleep fragmentation, sleep continuity, and total sleep time have all been linked to noise [24, 25]. There was no association between sleep duration and hourly minimum noise levels [86]. Again, it was also reported that there was no relation between sleep efficiency and mean noise levels, according to Missildine et al. [87].

Demographic characters	X^2	df	Asymp. Sig. (two-sided)
Gender	10.432	3	0.015
Age	10.889	6	0.092
Marital status	5.992	3	0.112
Qualification	28.342	9	0.001
Occupation	7.476	6	0.279

Table 9.
Relation between demographic character and drowsy.

Demographic characters	X^2	df	Asymp. Sig. (two-sided)
Gender	19.507	4	0.001
Age	21.261	8	0.006
Marital status	7.570	4	0.109
Qualification	7.787	12	0.802
Occupation	10.700	8	0.219

Table 10.
Relation between demographic character and relax.

Demographic characters	X^2	df	Asymp. Sig. (two-sided)
Gender	17.837	4	0.001
Age	21.817	8	0.005
Marital status	19.403	4	0.001
Qualification	21.981	12	0.038
Occupation	12.233	8	0.141

Table 11.
Relation between demographic character and sensitive.

Demographic characters	X^2	df	Asymp. Sig. (two-sided)
Gender	20.517	4	0.000
Age	9.797	8	0.280
Marital status	23.082	4	0.000
Qualification	13.892	12	0.308
Occupation	5.683	8	0.683

Table 12.
Relation between demographic character and Anexiety.

But, the result of the present study contradicts it and it shows that there is an association between sleep problems and noise level of the town ($p = 0.016$).

Table 13 depicts the results of ANNOVA analysis between noise annoyance and demographic characteristics. The table clearly depicts that there is an association between annoyance and gender of the study. However, there is no statistically significant link between other demographic factors and annoyance. There is a link between sex and anxiety ($p = 0.033$) as seen in **Table 14**. There is no direct relation between sensitivity with the demographic characters except marital status (**Table 15**). **Table 16** reveals that

Source	Type III sum of squares	Df	Mean square	F	Sig.
Corrected model	9.770[a]	5	1.954	2.289	0.049
Intercept	33.806	1	33.806	39.603	0.000
Age	0.111	1	0.111	0.130	0.719
Sex	6.623	1	6.623	7.759	0.006
Qualification	0.104	1	0.104	0.121	0.728
Marital status	0.107	1	0.107	0.126	0.724
Occupation	2.998	1	2.998	3.512	0.063
Error	122.923	144	0.854		
Total	2422.000	150			
Corrected total	132.693	149			

[a]*R squared = 0.074 (Adjusted R squared = 0.041).*

Table 13.
Analysis of ANNOVA between demographic characteristics and annoyance.

Source	Type III sum of squares	Df	Mean square	*F*	Sig.
Corrected model	11.910[a]	5	2.382	1.870	0.103
Intercept	27.567	1	27.567	21.642	0.000
Age	0.053	1	0.053	0.042	0.839
Sex	7.565	1	7.565	5.939	0.016
Qualification	1.247	1	1.247	0.979	0.324
Marital status	1.301	1	1.301	1.021	0.314
Occupation	1.171	1	1.171	0.919	0.339
Error	183.430	144	1.274		
Total	1909.000	150			
Corrected total	195.340	149			

[a]*R squared = 0.061 (Adjusted R squared = 0.028).*

Table 14.
Analysis of ANNOVA between demographic characteristics and anxiety.

Source	Type III sum of squares	df	Mean square	*F*	Sig.
Corrected model	32.004[a]	5	6.401	4.229	0.001
Intercept	5.506	1	5.506	3.638	0.058
Age	1.008	1	1.008	0.666	0.416
Sex	3.096	1	3.096	2.045	0.155
Qualification	0.891	1	0.891	0.589	0.444
Marital status	24.006	1	24.006	15.859	0.000
Occupation	0.787	1	0.787	0.520	0.472
Error	217.969	144	1.514		
Total	1930.000	150			
Corrected total	249.973	149			

[a]*R squared = 0.128 (Adjusted R squared = 0.098).*

Table 15.
Analysis of ANNOVA between demographic characteristics and sensitivity.

there is an association between relaxation and age (p = 0.008) and sex (p = 0.001) of the participants of the present study. **Table 17** shows the relation between annoyance and different environmental issues. This table clearly depicts that there is a strong association between relaxation, sensitivity, environmental noise, anxiety, irritation. Different vehicles are running on the main road of the town. During lock-down and unlock phases, ambulances are flowing from different areas of the town to the district hospital centre and also to the other clinics of the town. it has been reported that noise sensitivity—internal states that increase the chance of noise annoyance [88]—could alter the relationship between noise and health. Noise sensitivity has been linked to the beginning of depressed and psychological symptoms in adulthood. Higher morning saliva cortisol levels were linked to significant noise irritation and residing in high-noise locations in adolescents [89]. We did not have a way to gauge noise sensitivity or annoyance, so we could not assess its impact [90].

Source	Type III sum of squares	df	Mean square	*F*	Sig.
Corrected model	18.485[a]	5	3.697	3.841	0.003
Intercept	9.228	1	9.228	9.588	0.002
Age	6.870	1	6.870	7.138	0.008
Sex	10.157	1	10.157	10.553	0.001
Qualification	2.278	1	2.278	2.367	0.126
Marital status	.062	1	0.062	0.064	0.800
Occupation	1.534	1	1.534	1.594	0.209
Error	138.588	144	0.962		
Total	1055.000	150			
Corrected total	157.073	149			

[a]*R squared = 0.118 (Adjusted R squared = 0.087).*

Table 16.
Analysis of ANNOVA between demographic characteristics and relax.

Source	Type III sum of squares	df	Mean square	*F*	Sig.
Corrected model	67.409[a]	9	7.490	16.062	0.000
Intercept	0.906	1	0.906	1.942	0.166
Relax	6.410	1	6.410	13.746	0.000
Sensitive	8.560	1	8.560	18.356	0.000
Aware of noise pollution	0.010	1	0.010	0.021	0.884
Environmental noise	6.235	1	6.235	13.371	0.000
Hearing condition	0.643	1	0.643	1.379	0.242
Anexiety	4.360	1	4.360	9.350	0.003
Irritation	3.798	1	3.798	8.145	0.005
Depression	0.000	1	0.000	0.001	0.981
Health risk	8.296	1	8.296	17.790	0.000
Error	65.284	140	0.466		
Total	2422.000	150			
Corrected total	132.693	149			

[a]*R squared = 0.508 (Adjusted R squared = 0.476).*

Table 17.
Analysis of ANNOVA between annoyance and environmental factors.

The current research clearly shows that persons in the study locations are sensitive to noise levels based on their age. Respondents are employed in a variety of sub-urban work sites. They are subjected to various types of noise. They are irritated by the noise levels in the vicinity as a result of this. It is impossible to say that the level of noise in their workplace is the sole source of their annoyance, although it could be one of them.

During unlock phases, different offices are also opened in a regular and controlled manner. The running of vehicles on the road also growing accordingly and that may affect the public health in anyway. Different construction works also going on in many parts of the town and it may cause problem to the public of the town. Heavy vehicles carrying various raw materials are also moving on this road due to road building in various portions of the road. Vehicles are driven at all hours of the day and night. People of all ages are directly exposed to these levels of noise. This activity may exacerbate their sleeping problems.

4. Conclusion

Our findings may have been influenced by the fact that the noise level decreased due to the imposition of the nationwide lockdown and it then increase sharply due to the incoming of unlock phases. Still, the reported noise level of the town was beyond the permissible limit except lockdown phases in residential and silence zone. It was reported in the present study that there is a good association between different areas such as residential, commercial and silence zone with unlock and before lock down phase of the town. In case of monthly noise level variation with different phases of the lockdown situation there is also good association between them and is presented in the **Table 5**. Finally, studies have demonstrated that the relationship between noise and health differs depending on sex, health status, and other factors but we lacked the sample size to evaluate the relationship by subgroup. Longitudinal designs, enhanced exposure assessment, and objective sleep assessments of whether particular subgroups of teenagers are more susceptible to the potential negative effects of environmental noise, should be prioritised in future investigations. Direct regulation of noise sources as well as changes to the built environment are two public health techniques for reducing noise exposure [21, 91]. We were unable to demonstrate a temporal relationship between exposure and outcome since the study was cross-sectional. Future research may want to utilise objective of audiometry test to test the exactness of the hearing quality of the respondents of the town.

Acknowledgements

The authors are very much thankful to Indrajit Patra and Pravat Kumar Mandal for their support in monitoring the noise levels.

Conflicts of interest

The authors declare that they have no conflicts of interest with regard to the content of this report.

Author details

Bijay Kumar Swain[1*], Chidananda Prasad Das[2] and Shreerup Goswami[3]

1 DIET, Bhadrak, Odisha, India

2 Environmental Science Program, Department of Chemistry, ITER, S 'O' A Deemed to be University, Bhubaneswar, Odisha, India

3 PG Department of Geology, Utkal University, Vani Vihar, Bhubaneswar, Odisha, India

*Address all correspondence to: bkswain.dr@gmail.com; swainbijay0@gmail.com

References

[1] Kerns E, Masterson EA, Themann CL, Calvert GM. Cardiovascular conditions, hearing difficulty, and occupational noise exposure within US industries and occupations. American Journal of Industrial Medicine. 2018;**61**(6):477-491. DOI: 10.1002/ajim.22833

[2] Licitra G, Ascari E, Fredianelli L. Prioritizing process in action plans: A review of approaches. Current Pollution Reports. 2017;**3**:151-161. DOI: 10.1007/s40726-017-0057-5

[3] Fiedler K, Wilhelm M. Hygiene/Preventive Medicine/Environmental Medicine. Bremen, London, Boston: UNI-MED Verlag AG. (In German); 2011

[4] Tak S, Davis RR, Calvert GM. Exposure to hazardous workplace noise and use of hearing protection devices among US workers—NHANES, 1999-2004. American Journal of Industrial Medicine. 2009;**52**(5):358-371. DOI: 10.1002/ajim.20690

[5] Das CP, Swain BK, Goswami S, Das M. Prediction of traffic noise induced annoyance: A two-staged SEM-Artificial Neural Network approach. Transportation Research Part D. 2021;**2021**(100):1-20. DOI: 10.1016/j.trd.2021.103055

[6] Goswami S, Swain BK. Environmental noise in India: A review. Current Pollution Reports. 2017;**3**(3):220-229

[7] Engel MS, Zannin PHT. Noise assessment of the area of a redesigned urban expressway based on noise measurements, noise maps and noise perception interviews. Noise Control Engineering Journal. 2017;**65**:590-610. DOI: 10.3397/1/376572

[8] Méline J, Van Hulst A, Thomas F, Karusisi N, Chaix B. Transportation noise and annoyance related to road traffic in the French RECORD study. International Journal of Health Geographics. 2013;**12**(1):1-13

[9] Swain BK, Goswami S. Acoustic environment in the bus: An empirical study. Pollution. 2018;**4**(2):327-333

[10] Swain BK, Goswami S. Soundscapes of urban parks in and around Bhubaneswar and Puri, Odisha, India: A comparative study. Pollution. 2018;**4**(1):93-101

[11] Swain BK, Goswami S, Das M. Appraisal and assessment of noise level during the Dussehera festival: A case study of Balasore, India. International Journal of Earth Sciences and Engineering. 2013;**6**(2):375-380

[12] Swain BK, Goswami S, Panda SK. Road traffic noise assessment and modeling in Bhubaneswar, India: A comparative and comprehensive monitoring study. International Journal of Earth Sciences and Engineering. 2012;**5**(1):1358-1370

[13] Babisch W. The noise/stress concept, risk assessment and research needs. Noise & Health. 2002;**4**(16):1-11

[14] Jensen HAR, Rasmussen B, Ekholm O. Neighbour noise annoyance is associated with various mental and physical health symptoms: Results from a nationwide study among individuals living in multi-storey housing. BMC Public Health. 2019;**19**:1-10

[15] Alkhalawi E, Orban E, Schramm S, Katsarava Z, Hoffmann B, Moebus S. Residential traffic noise exposure and headaches: Results from the population-based heinz nixdorf recall study. Noise & Health. 2021;**23**:1-10

[16] Sygna K, Aasvang GM, Aamodt G, Oftedal B, Krog NH. Road traffic noise, sleep and mental health. Environmental Research. 2014;**131**:17-24

[17] Franssen EA, van Wiechen CM, Nagelkerke NJ, Lebret E. Aircraft noise around a large international airport and its impact on general health and medication use. Occupational and Environmental Medicine. 2004;**61**(5): 405-413

[18] Jarup L, Babisch W, Houthuijs D, Pershagen G, Katsouyanni K, Cadum E, et al. Hypertension and exposure to noise near airports: The HYENA study. Environmental Health Perspectives. 2008;**116**(3):329-333

[19] Sorensen M, Andersen ZJ, Nordsborg RB, Becker T, Tjonneland A, Overvad K, et al. Long-term exposure to road traffic noise and incident diabetes: A cohort study. Environmental Health Perspectives. 2013;**121**(2):217-222

[20] Nitschke M, Tucker G, Simon DL, Hansen AL, Pisaniello DL. The link between noise perception and quality of life in South Australia. Noise & Health. 2014;**16**:137-142

[21] Rudolph KE, Shev A, Paksarian D, Merikangas KR, Mennitt DJ, James P, et al. Environmental noise and sleep and mental health outcomes in a nationally representative sample of urban US adolescents. Environmental Epidemiology. 2019;**3**:1-9

[22] Basner M, McGuire S. WHO environmental noise guidelines for the European region: A systematic review on environmental noise and effects on sleep. International Journal of Environmental Research and Public Health. 2018;**15**:519

[23] Weyde KV, Krog NH, Oftedal B, et al. Nocturnal road traffic noise exposure and children's sleep duration and sleep problems. International Journal of Environmental Research and Public Health. 2017;**14**:491

[24] Basner M, Babisch W, Davis A, Brink M, Clark C, Janssen S, et al. Auditory and non-auditory effects of noise on health. Lancet. 2014;**383**: 1325-1332

[25] Muzet A. Environmental noise, sleep and health. Sleep Medicine Reviews. 2007;**11**(2):135-142

[26] Miedema HM, Vos H. Associations between self-reported sleep disturbance and environmental noise based on reanalyses of pooled data from 24 studies. Behavioral Sleep Medicine. 2007;**5**(1):1-20

[27] Oh M, Shin K, Kim K, Shin J. Influence of noise exposure on cardio cerebrovascular disease in Korea. Science of the Total Environment. 2019;**651**:1867-1876

[28] Orban E, McDonald K, Sutcliffe R, et al. Residential road traffic noise and high depressive symptoms after five years of follow-up: Results from the Heinz Nixdorf recall study. Environmental Health Perspectives. 2016;**124**:578

[29] van Kempen E, Babisch W. The quantitative relationship between road traffic noise and hypertension: A meta-analysis. Journal of Hypertension. 2012;**30**:1075, 12-1086

[30] Babisch W, Kamp I. Exposure–response relationship of the association between aircraft noise and the risk of hypertension. Noise & Health. 2009;**11**(44):161-168

[31] Munzel T, Gori T, Babisch W, Basner M. Cardiovascular effects of environmental noise exposure. European Heart Journal. 2014;**35**:829-836

[32] Abo-Qudais S, Alhiary A. Effect of distance from road intersection on developed traffic noise levels. Canadian Journal of Civil Engineering. 2004;**31**(4):533-538

[33] Oyedepo SO, Adeyemi GA, Olawole OC, Ohijeagbon OI, Fagbemi OK, Solomon R, et al. A GIS-based method for assessment and mapping of noise pollution in Ota metropolis, Nigeria. MethodsX. 2019;**6**:447-457

[34] Pretzsch A, Seidler A, Hegewald J. Health effects of occupational noise. Current Pollution Reports. 2021;7: 344-358

[35] Mishra A, Das S, Singh D, Maurya ASK. Effect of COVID-19 lockdown on noise pollution levels in an Indian city: A case study of Kanpur. Environmental Science and Pollution Research. 2021;**28**:46007-46019. DOI: 10.1007/s11356-021-13872-z

[36] Zannin PHT, do Valle F, do Nascimento EO. Assessment of noise pollution along two Main avenues in Curitiba, Brazil. Open Journal of Acoustics. 2019; 9: 26-38. 10.4236/ oja.2019.92003.

[37] CPCB. Ambient Air Quality in Respect of Noise. New Delhi: Central Pollution Control Board; 2000 Schedule-Part II, Sec. 3

[38] Goswami S. Road traffic noise: A case study of Balasore Town, Orissa, India. International Journal of Environmental Research. 2009;**3**:309-316

[39] Goswami S. The soundscape of Bhadrak town, India: An analysis from road traffic noise perspective. Asian Journal of Water, Environment and Pollution. 2011;**8**:85-91

[40] Goswami S, Nayak SK, Pradhan AC, Dey SK. A study on traffic noise of two campuses of university, Balasore, India. Journal of Environmental Biology. 2011;**32**:105-109

[41] Goswami S, Swain BK, Mohapatra HP, Bal KK. A preliminary assessment of noise level during Deepawali festival in Balasore, India. Journal of Environmental Biology. 2013;**34**(6):981-984

[42] Goswami S, Swain BK, Panda SK. Assessment, analysis and appraisal of road traffic noise pollution in Rourkela city, India. Journal of Environmental Biology. 2013;**34**(5):891-895

[43] Pradhan A, Swain BK, Goswami S. Road traffic noise assessment and modeling of Sambalpur City, India: A comprehensive, comparative and complete study. Journal of Ecophysiology and Occupational Health. 2012;**12**:51-63

[44] Pradhan AC, Swain BK, Goswami S. Measurements and model calibration of traffic noise pollution of an industrial and the Intermediate City of India. The Ecoscan. 2012;**1**:1-4

[45] Goswami S, Swain BK. Soundscape of Balasore City, India: A study on urban noise and community response. Journal of Acoustical Society of India. 2011;**38**(2):59-71

[46] Goswami S, Swain BK. Soundscape of Baripada, India: An appraisal and evaluation from urban noise perspective. The Ecoscan Special Issue. 2013;**3**:29-34

[47] Sahu SK, Swain BK, Das M, Goswami S. Data of monitored noise and predictive models in and around Rayagada, Odisha, India: A comparative study. International Journal for Innovative Research in Science & Technology. 2014;**5**:39-47

[48] Swain BK, Goswami S, Das M. Assessment of highway noise and

predictive models along NH-316. Asian Journal of Water, Environment and Pollution. 2016;**13**(2):97-105

[49] Swain BK, Panda SK, Goswami S. Dynamics of road traffic noise in Bhadrak City, India. Journal of Environmental Biology. 2012;**33**: 1087-1092

[50] Swain BK, Goswami S. Data of monitored highway noise and predictive models: A relative and inclusive case study. International Journal of Earth Sciences and Engineering. 2013;**6**(5):1079-1085

[51] Swain BK, Goswami S. Integration and comparison of assessment and modeling of road traffic noise in Baripada town, India. International Journal of Energy and Environment. 2013;**4**(2):303-310

[52] Swain BK, Goswami S. Analysis and appraisal of urban road traffic noise of the City of Cuttack, India. Pakistan Journal of Scientific and Industrial Research Series A: Physical Sciences. 2014;**57**(1):10-19

[53] Kalawapudi K, Singh T, Dey J, Vijay R, Kumar R. Noise pollution in Mumbai metropolitan region (MMR): An emerging environmental threat. Environmental Monitoring and Assessment. 2020;**192**(2):1-20

[54] Thakre C, Laxmi V, Vijay R, Killedar DJ, Kumar R. Traffic noise prediction model of an Indian road: An increased scenario of vehicles and honking. Environmental Science and Pollution Research. 2020;**27**(30):38311-38320

[55] Swain BK, Goswami S, Das M. A preliminary study on assessment of noise levels in Indian offices: A case study. Asian Journal of Water, Environment and Pollution. 2014;**11**(4):39-44

[56] Goswami S, Swain BK. Preliminary information on noise pollution in commercial banks of Balasore, India. Journal of Environmental Biology. 2012;**33**(6):999-1002

[57] Swain BK, Goswami S. A study on noise in Indian banks: An impugnation in the developing countries. Pakistan Journal of Scientific and Industrial Research Series A: Physical Sciences. 2014;**57**(2):103-108

[58] Mohapatra H, Goswami S. Assessment and analysis of noise levels in and around Ib River coalfield, Orissa, India. Journal of Environmental Biology. 2012;**33**:649-655

[59] Mohapatra H, Goswami S. Assessment of noise levels in various residential, commercial and industrial places in and around Belpahar and Brajrajnagar, Orissa, India. Asian Journal of Water Environment and Pollution. 2012;**9**:73-78

[60] Goswami S, Swain BK. Occupational exposure in stone crusher industry with special reference to noise: A pragmatic appraisal. Journal of Acoustical Society of India. 2012;**39**(2):70-81

[61] Swain BK, Goswami S, Tripathy JK. Stone crushers induced noise at and around Mitrapur, Balasore, India. Anwesa. 2011;**6**:12-16

[62] Zambon G, Confalonieri C, Angelini F, Benocci R. Effects of COVID-19 outbreak on the sound environment of the city of Milan, Italy. Noise Mapping. 2021;**8**:116-128. DOI: 10.1515/noise-2021-0009

[63] Terry C, Rothendler M, Zipf L, Dietze MC, Primack RB. Effects of the COVID-19 pandemic on noise pollution in three protected areas in metropolitan Boston (USA). Biological Conservation. 2021;**256**:109039

[64] Hornberg J, Haselhoff T, Lawrence BT, Fischer JL, Ahmed S, Gruehn D, et al. Impact of the COVID-19 lockdown measures on noise levels in urban areas—A pre/during comparison of long-term sound pressure measurements in the Ruhr area, Germany. International Journal of Environmental Research and Public Health. 2021;**18**:4653. DOI: 10.3390/ijerph18094653

[65] Caniato M, Bettarello F, Gasparella A. Indoor and outdoor noise changes due to the COVID-19 lockdown and their effects on individuals' expectations and preferences. Scientific Reports. 2021;**11**:16533. DOI: 10.1038/s41598-021-96098-w

[66] Alías F, Alsina-Pagès RM. Effects of COVID-19 lockdown in Milan urban and Rome suburban acoustic environments: Anomalous noise events and intermittency ratio. The Journal of the Acoustical Society of America. 2022;**151**:1676. DOI: 10.1121/10.0009783

[67] Laxmi V, Dey J, Kalawapudi K, Vijay R, Kumar R. An innovative approach of urban noise monitoring using cycle in Nagpur, India. Environmental Science and Pollution Research. 2019;**26**:36812-36819. DOI: 10.1007/s11356-019-06817-0

[68] Tiwari K, Jha SK, Tandel BN. GIS based approach for noise mapping of urban road traffic. NAP-2017. 2017. S.V.N.I.T. Available from: https://www.researchgate.net/profile/Kartikey-Tiwari/publication/319261157_GIS_Based_Approach_for_Noise_Mapping_of_Urban_Road_Traffic/links/599eb4110f7e9b892bb8fad4/GIS-Based-Approach-for-Noise-Mapping-of-Urban-Road-Traffic.pdf

[69] De SK, Swain BK, Goswami S, Das M. Adaptive noise risk modeling: Fuzzy logic approach. Systems Science & Control Engineering. 2017;**2017**(5):129-141

[70] Oyedepo SO. Development of noise map for Ilorin metropolis, Nigeria. International Journal of Environmental Studies. 2013;**1**:12. DOI: 10.1080/00207233.2013.813716

[71] Vianna KMP, Cardoso MRA, Rodrigues RMC. Noise pollution and annoyance: An urban soundscapes study. Noise & Health. 2015;**17**:125-133

[72] Weinstein ND. Individual differences in reactions to noise: A longitudinal study in a college dormitory. Journal of Applied Psychology. 1978;**63**:458-466

[73] Allimohammadi I, Nassiri P, Azkhosh M, Hoseini M. Factors affecting road traffic noise annoyance among white-collar employees working in Tehran, Iran. Journal of Environmental Health Science & Engineering. 2010;**7**(1):25-34

[74] World Health Organization. Occupational and Community Noise. WHO-OMS. 2005. Available from: http://www.who.int.inffs/en/fact.html

[75] World Health Organization. Environmental noise guidelines for the European region. 2018. Available at: https://www.euro.who.int/en/publications/abstracts/environmental-noise-guidelines-for-theeuropean-region-2018

[76] WHO. Night Noise Guidelines for Europe. Copenhagen Ø, Denmark: World Health Organization-WHO Regional Office for Europe; 2009

[77] Berguland B, Lindvall T, Schwela D. New who guidelines for community noise. Noise & Vibration Worldwide. 2000;**31**(4):24-29

[78] Krishna Murthy V, Majumdar AK, Khanal SN, Subedi DP. Assessment of

traffic noise pollution in Banepa, a semi urban town of Nepal. Kath. Uni. J. Sci. Engg. Tech. 2007;**1**:1-9

[79] DGUV (Deutsche Gesetzliche Unfallversicherung e.V.). DGUV Statistiken für die Praxis 2019—Aktuelle Zahlen und Zeitreihen aus der Deutschen Gesetzlichen Unfallversicherung. Berlin; 2019 Contract No.: Webcode: p021547. Available from: https://publikationen.dguv.de/widgets/pdf/download/article/3903

[80] Kieffer C. Reporting of Occupational Diseases: Issues and Good Practices in Five European Countries. Paris: EUROGIP; 2015 Contract No.: Eurogip-102/E

[81] Koskela K. Noise-induced hearing loss is still the most common confirmed occupational disease. In: Work-Life Knowledge Service. Finnish Institute of Occupational Health; 2020. Available from: https://www.worklifedata.fi/#/en/articles/analysisOccupationalDiseases-10-2020

[82] Öhrström E, Hadzibajramovic E, Holmes M, Svensson H. Effects of road traffic noise on sleep: Studies on children and adults. Journal of Environmental Psychology. 2006;**26**:116-126

[83] Okokon EO, Yli-Tuomi T, Turunen AW, Tiittanen P, Juutilainen J, Lanki T. Traffic noise, noise annoyance and psychotropic medication use. Environment International. 2018;**119**:287-294. DOI: 10.1016/j.envint.2018.06.034

[84] Guski R, Schreckenberg D, Schuemer R. WHO environmental noise guidelines for the European region: A systematic review on environmental noise and annoyance. International Journal of Environmental Research and Public Health. 2017;**14**:1-39. DOI: 10.3390/ijerph14121539

[85] Park SH, Lee PJ, Jeong JH. Effects of noise sensitivity on psychophysiological responses to building noise. Building and Environment. 2018;**136**:302-311. DOI: 10.1016/j.buildenv.2018.03.061

[86] Adachi M, Staisiunas PG, Knutson KL, Beveridge C, Meltzer DO, Arora VM. Perceived control and sleep in hospitalized older adults: A sound hypothesis? Journal of Hospital Medicine. 2013;**8**:184-190

[87] Missildine K, Bergstrom N, Meininger J, Richards K, Foreman MD. Sleep in hospitalized elders: A pilot study. Geriatric Nursing. 2010;**31**:263-271

[88] Job RF. Noise sensitivity as a factor influencing human reaction to noise. Noise & Health. 1999;**1**:57-68

[89] Stansfeld SA, Shipley M. Noise sensitivity and future risk of illness and mortality. Science of the Total Environment. 2015;**520**:114-119

[90] Wallas A, Eriksson C, Gruzieva O, et al. Road traffic noise and determinants of saliva cortisol levels among adolescents. International Journal of Hygiene and Environmental Health. 2018;**221**:276-282

[91] Hammer MS, Swinburn TK, Neitzel RL. Environmental noise pollution in the United States: Developing an effective public health response. Environmental Health Perspectives. 2014;**122**:115

Chapter 5

Evacuation Guidance Assistance System Using Emitting Sound

Tetsuya Miyoshi

Abstract

The goal of our research is developing an evacuation guidance system that emits sounds on a set of loudspeakers in a spatial sequence to achieve rapid evacuation in emergency situations. For this goal, we evaluate the auditory recognition properties of emitting sounds and their ability to make evacuees follow the movement of sound stimuli in this chapter. We conduct three experiments to assess the proposed evacuation guidance method. The first and the second ones are done to investigate the recognition properties of the positions and directions of the emitting sound stimuli. The third one is done to evaluate the ability to guide evacuees to exit using the emitting sound in the spatial sequence. In the first experiment, we consider whether the four factors related to the sound-emitting method affect the identification of the emitting sound stimuli. Additionally, we investigate the patterns of emitted sounds that can easily recognize their position and direction. In the third experiment, we consider whether the evacuee can follow the emitting sound on a set of loudspeakers in spatial sequence. Moreover, it is discussed that the proposed guidance system provided a more detailed evacuation route for evacuees.

Keywords: evacuation guidance system, sound localization, auditory recognition, human behavior for acoustic stimuli, spatial sequence of sound

1. Introduction

In the event of a disaster, such as fire outbreak or earthquake, prompt evacuation to a safe zone is important and essential for ensuring the safety of evacuees. Facility managers are obliged to install fire extinguishing equipment, sufficient emergency exit and guide lights to make evacuee find proper path to the exit by the Japanese Fire Service Law, even if they are not familiar with the layout of the area [1]. Recently, several urban commercial premises have become huge and complex in order to provide efficiency and convenience. Evacuees are not able to intuitively find out evacuation routes in these buildings and structures without proper guidance to exits due to their spatial complexity. Active evacuation guidance systems have been developed to control different light and acoustic stimuli that provide evacuation guidance information to evacuees to construct a safe and secure evacuation environment [2–6].

Conventional evacuation guidance systems in buildings and structures are not designed to react to changes in situations, such as collapses or other disturbances.

Evacuation guidance systems must be able to autonomously determine which evacuation routes have not been damaged by the disaster to achieve rapid evacuation in emergency situations. An autonomous route-detection system, in which several smoke and heat sensors could be placed at key points in the objective area to determine the evacuation route based on the overall condition, was proposed [7]. An evacuation route guidance system that considers evacuees' current location and building safety using a smart building-sensor network and is able to recommend the best evacuation route for each localized evacuee through their mobile terminals was proposed and evaluated [8]. A method for determining evacuation routes has been proposed that uses location information from mobile devices to determine effective routes [9, 10].

It is crucial to guide the evacuees along the determined evacuation route to effectively use route-detection systems in disasters. However, the evacuation and pathway guide lights in the conventional evacuation guidance were not sufficient to lead the evacuees to the relevant routes adapted to the situation. Therefore, a system is required to guide evacuees flexibly in disasters. Several systems that help evacuees select the route to appropriately exit using light and sound stimuli have been proposed.

Several previous studies have researched active evacuation guidance systems that utilize the precedence effect (Haas effect) to help evacuee realize evacuation direction. It is a psychological feature in hearing acrostic stimuli. When two identical sounds are presented in close succession, the spatial location of the auditory stimulus is dominated by the first arriving sound [11]. Additionally, the implementation of sound equipment (such as loudspeakers and signal processors) and sound-stimuli presentations in evacuation guidance systems have been standardized by the Japan Lighting Manufactures Association (JLMA), such that evacuees can correctly identify evacuation routes [2]. Furthermore, an improved evacuation system utilizing the precedence effect, in which loudspeakers were set beside a wall in a passageway to avoid the disappearance of the precedence effect of an audio signal, was proposed [12, 13].

The evacuation guidance systems using the precedence effect able to lead evacuees to one or two exits predefined as an emergency exit, however, they provide only the direction to the exit and not a detailed evacuation path to the exit using acoustic stimulus. Therefore, the evacuee must discover an evacuation route to the exit using acoustic stimulus even if the guidance using the precedence effect to the exit were provided. If there were several obstacles in the current place, it might not be easy to avoid damaged passageways and a fire outbreak caused by a disaster.

Passengers and crew may quickly lose situational awareness in a smoke-filled cabin of an aircraft. The European Union Aviation Safety Agency (EASA) and Federal Aviation Administration (FAA) regulations stipulate requirements for emergency floor-path illumination in all aircraft to achieve faster evacuation. Thus, guiding pathways to exits makes sense in situations where vision does not work well. However, there are few studies that provide a pathway to exits using emitting acoustic stimuli sequentially. Therefore, in this study, we propose a new active evacuation guidance system using acoustic cues, in which guidance-sound stimuli are sequentially emitted along an evacuation path instead of relying on the precedence effect [6].

This study's objective is to develop an active evacuation guidance system to direct evacuees along an evacuation route by sequentially emitting sound stimuli. In the first stage of the study, we analyzed participants' capacity to identify sound stimuli emitted through a set of loudspeakers. We conducted experiments to investigate the recognition properties of the position and direction of the emitting sound, in which four factors, such as the stimulus type and emission-time interval might affect their capacity to identify the stimuli. Additionally, the identification performance of the evacuee for the emitting sequences along the straight and bent lines was considered. Subsequently, we considered whether the evacuee could follow the emitting sound on a set of loudspeakers in sequence. Furthermore, we demonstrated that the proposed guidance system using the sound provided a more detailed evacuation route for evacuees.

The remainder of this paper is organized as follows. Section 2 presents the advantages of the proposed evacuation guidance system that emits sound stimuli sequentially. Section 3 summarizes properties of auditory recognition for the emitting sound stimuli based on our previous research [6]. Section 4 describes the subjects' ability to follow the sequence of the emitting sound based on experiments and discusses the practicality and feasibility of the proposed evacuation guidance systems. Finally, the usefulness of a sound-based guidance system proposed in this paper is summarized in Section 5.

2. Evacuation guidance systems using emitting sound stimuli

First, an overview and advantages of evacuation guidance systems using emitting sound sources on a set of loudspeakers are described. Conventional guidance systems assume that evacuees can determine their evacuation route based on guide lights in an emergency situation. However, it might be difficult for them to find the relevant or correct route if they are not familiar with the spatial location. Therefore, there are two types of evacuation guidance systems that use acoustic stimuli to indicate the evacuation direction. In this study, we propose a method that uses an emitting sound source on a loudspeaker along routes to guide evacuees. If several loudspeakers are placed in the objective area, the sound sources could be sequentially emitted from one loudspeaker to another. People would be able to recognize them as the stream of sound in the evacuation direction. The recognition of the direction of the sound stream is generalized by the sequence of sound localization for a single-sound source.

In shopping centers, a wide floor is occupied by display cases of goods, which can be an obstacle to evacuation in the event of a disaster. The proposed new evacuation guidance system would be able to provide a pathway to avoid these obstacles because it shows the sequence of the route to exit by emitting sound. This study describes the feasibility and performance of the new evacuation guidance system using the emitting acoustic sound stimuli.

3. Identification of emitting sound patterns in sequences

In this section, as a first step towards developing the guidance system utilizing a sound sequence emitting a sound stimulus to induce people to exit, two experiments to test whether people can identify the sound sequence is described with reference to our previous research [6]. The actual evacuation routes include a variety of patterns, so the two sequence patterns of sounds, straight-line and right-angle patterns were evaluated in the following subsection.

3.1 Identification of emitting sound in straight-line sequences

3.1.1 Factors considered in experiments

First, this study assesses individuals' capacity to identify a spatial sequence of sound stimuli and examines the factors that influence individuals' identification performance. Therefore, two experiments were conducted for ten healthy male students (20–21 years old) of Hannan University with no abnormal hearing diagnosed during their annual medical examination. They participated in this experiment without remuneration. Experimenter and all participants provided informed consent before these experiments.

The identification procedure of sound sequence requires the participants to continuously recognize sound localization for sequentially emitting sound through the loudspeakers. The sound localization is a listener's ability to identify the location of a detected sound in direction and distance. Various factors that cause changes in the pressure and sound wave frequency affect individuals' sound localization performance. Therefore, it is expedient to consider the influence of emission speed (emission-time interval) and the distance between loudspeakers to assess the subjects' identification performance. Furthermore, considering the preceding effect, the position of the subject relative to the loudspeaker should be considered a factor affecting the identification of the sound stimulus. Thus, the stimulus type, emission-time interval, distance between loudspeakers, and subject's position were considered as experimental factors in the first experiment.

Previous studies on emergency-alert sounds reported that stimuli containing a wide range of frequencies are more likely to be recognized than those with a single frequency [12]. Furthermore, swept-sound stimuli from low to high frequencies can be recognized more easily during evacuation guidance procedures. Therefore, in this experiment, the phrase "Here is an emergency exit," spoken by a female voice in Japanese because the human voice is an acoustic stimulus with many frequencies superimposed on it. It also was used in previous studies on evacuation guidance procedures using the preceding effect [4, 5].

As a result of the above considerations, the two types of acoustic stimulus were set to experimental factor, the female voice and a sound that linearly changed from 500 to 1000 Hz in 1 s (swept-sound). The distance between the loudspeakers was set to two levels, 3 and 5 m, while the emission-time intervals were set to 1 and 0.5 s. The longer the time interval, the faster the sound moved. The subjects were instructed to stand in one of two fixed places—just below a loudspeaker in the center of the grid or between loudspeakers.

3.1.2 Methodology of experiment 1

Our proposed evacuation guidance system help evacuee realize the path to the exit using sequence of acoustic stimuli emitted thought several loudspeakers which were arranged on the ceiling of buildings. In the first experiment loudspeakers were arranged in a 5 × 5 grid 4 m above from floor level in the gymnasium of Hanna University. The experimental environment and the arrangement of the loudspeakers are illustrated in **Figures 1** and **2**. The numbers rounded with squares in **Figure 2** are the index of loudspeakers.

During the experiment, the subjects stood in one of two possible positions, just below loudspeaker 13 or between loudspeakers 13 and 17. The loudspeakers were a

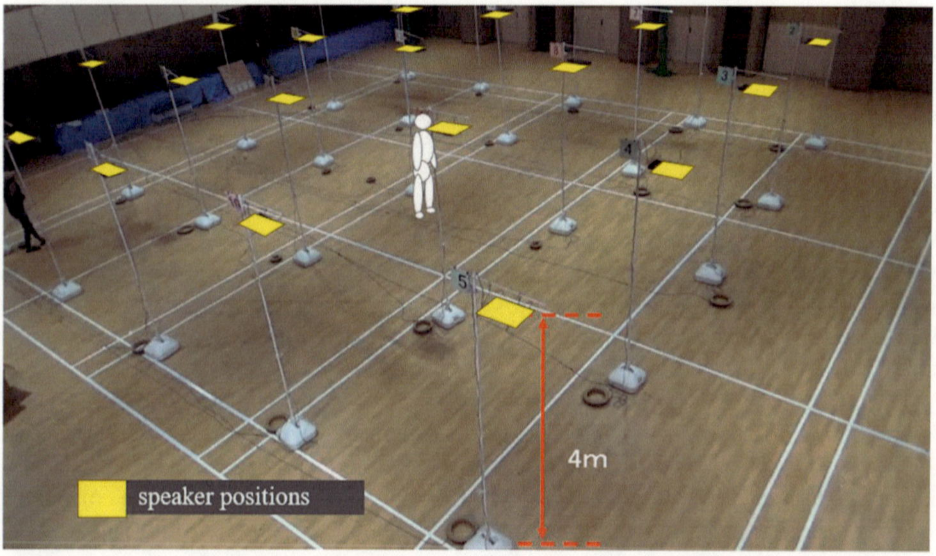

Figure 1.
Experimental environment and the arrangement of the loudspeakers.

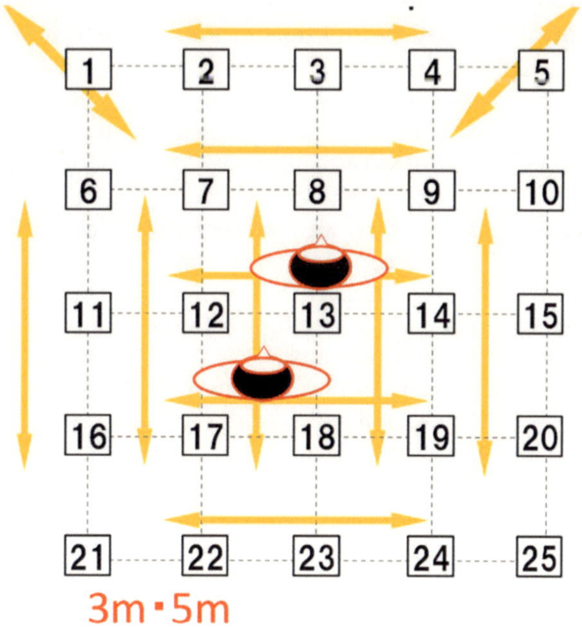

Figure 2.
Arrangement of the loudspeakers, subjects' position, and sequential pattern of sound stimuli in the first experiment.

capacitor-type flat speakers with stronger directionality than conventional dynamic speakers. We used a switching device with a small controller (Arduino Uno) that could be controlled by the software to emit the sound stimulus on all 25 loudspeakers in a sequence for specific time intervals. The sound stimulus level was set for each stimulus type (voice or swept-sound), such that the A-weighted noise level was

80 dB at a position 1 m from the loudspeaker. The noise level was measured using an integrated average-type sound-level meter (LA-1441, Ono Sokki Co., Ltd.).

In the first experiment, the sound stimulus (the voice or swept-sound) was emitted sequentially through five loudspeakers. For example, it is the sequence from loudspeakers 1–21 in the order of 1, 6, 11, 16, 21 in a straight line shown by an arrow in **Figure 2**. All 12 distinct sequence patterns of the sound stimuli (including five row-wise ones (left-to-right or right-to-left), five column-wise ones (front-to-back or back-to-front), and two diagonal ones (front-to-back or back-to-front) are shown by an arrow in **Figure 2**. Considering that the sequences were emitted in both ascending and descending order, there were totally 24 sequence patterns. For example, the emitting sequence in ascending order was set to five straight line patterns: numbers 1–5, 6–10, 11–15, 16–20, and 21–25.

The subjects were instructed to listen to the sequence at a specific position (as mentioned earlier, position just below loudspeaker 13 or the other) and identify the sequence pattern as quickly as possible. We conducted a trial for each subject to listen to the sound stimuli and answer which patterns were emitted before conducting the first experiment. The first experiment, in which the 24 sequence patterns were emitted randomly in each trial under different conditions (combining the four factors) was then conducted. The sound stimulus continued until the subjects returned their answer. Sixteen trials under each experimental condition combining the four factors were conducted for each subject repeatedly because four two-level experimental factors were considered in this experiment. An experimenter recorded the participants' response time and the accuracy of their answers (accuracy rate).

3.1.3 Experimental results and discussion for experiment 1

3.1.3.1 Experimental results regarding each factor

Figure 3 shows the mean accuracy rates and response times of identification of emitting sequence regarding four experimental factors. We conducted a three-way analysis of variance (ANOVA) to compare the mean-accuracy rates and response times, considering three within-subject factors: stimulus type, emission-time interval, and the distance between loudspeakers.

In the ANOVA results for accuracy rate, significant differences were observed for both stimulus type ($p = 0.01$) and emission interval ($p < 0.01$). Similarly, in the ANOVA for the response time, a significant difference was observed in the stimuli type ($p < 0.001$). Furthermore, an interaction was observed between the emission interval and the distance between the loudspeakers ($p < 0.001$). Thus, the results of ANOVA for accuracy rate and response time indicate that the sequence of voice stimuli emitted at 1.0 s interval is better than the other sequence.

The influence of four factors on the identification of sequence pattern of sound stimuli was evaluated through the experiment. The overall identification rate was over 80% across 24 sequence patterns. Comparing four factors affecting the accuracy and response time, the significant effects with respect to the type of sound source and the emission-time interval was confirmed. The type of sound source had a particularly strong effect for the results. The accuracy rate of identification was higher, and the response time was shorter when the voice sound emitted as acoustic stimuli than the swept-sound. These results suggest that the use of voice rather than swept-sound as a sound source enables the correct recognition of the direction of guidance.

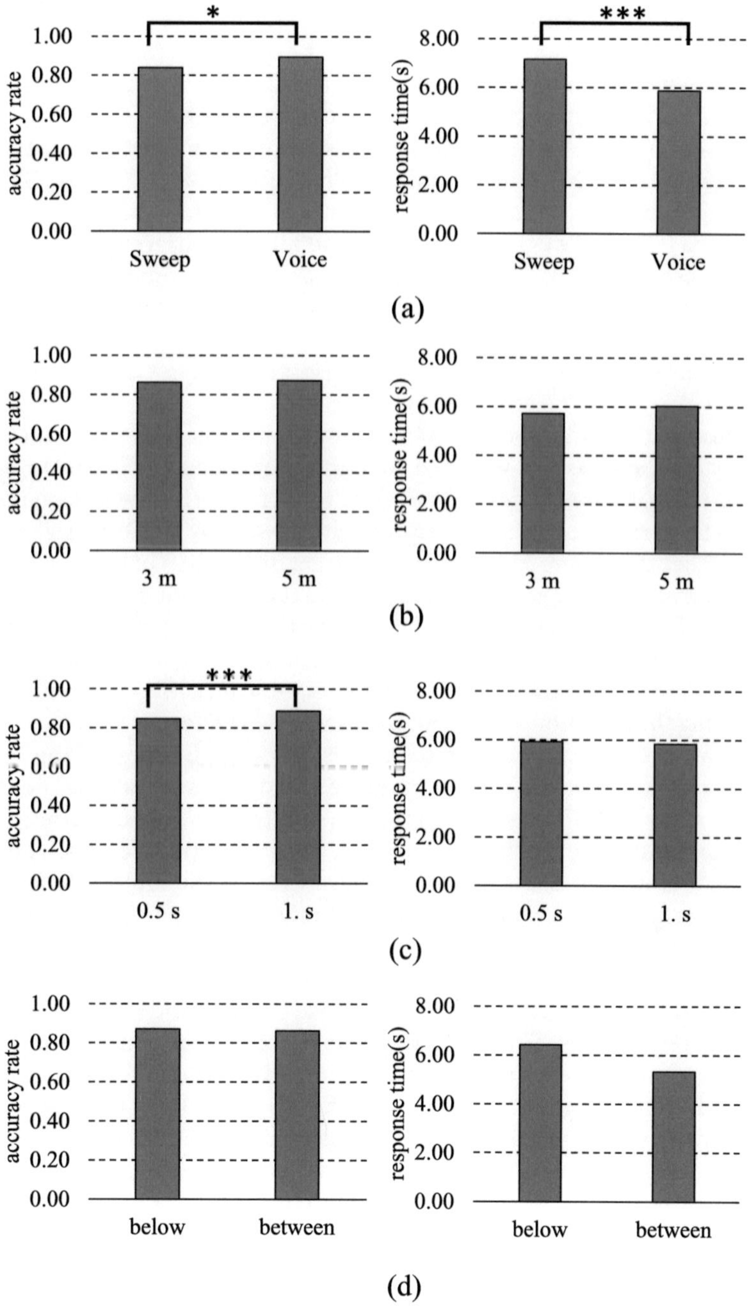

Figure 3.
Mean accuracy rates and response times regarding experimental factors. (From Miyoshi [6]). (a) Mean accuracy rates and response times regarding stimulus type. (b) Mean accuracy rates and response times regarding the distance between loudspeakers. (c) accuracy rates and response times regarding interval between sound stimuli. (d) Mean accuracy rates and response times regarding subject's standing point. $p < 0.001$: ***, $p < 0.01$: **, $p < 0.05$: *.

Aoki conducted the sound localization experiments for middle-aged and elderly subjects using multiple sound sources including voice as acoustic stimuli. As the results it was reported that the incorrect response rate was lowest, and the reaction time was shortest when vocal stimuli were used [13]. Thus, we could easily perceive the vocal phrase and identify its localization. Our experiment task was the identification of the sequence of emitting sound source, and the performance of task become higher for voice stimuli due to the ease of sound location for it.

The differences in the accuracy rates and response times classed according to the different emission-time intervals (1 and 0.5 s) are summarized in **Figure 3b**). The mean-accuracy rate for both sound stimuli emitted in the interval 1 s was higher than in interval 0.5 s, although there was not a significant difference in response time between both intervals. These results indicate that the stimulus sequence can be identified more easily when the emission-time interval is 1 s.

3.1.3.2 Experimental results with respect to identification of sequence patterns

The accuracy rates and response times of the subjects were compared among sequence patterns, such as row-wise and column-wise sequences. **Figure 4a** and **b** illustrate the mean accuracy rates and response times for the row-wise sequence patterns and the left-right direction, respectively. A two-way ANOVA was conducted to detect whether there were statistically significant differences in the mean scores regarding the five row-wise sequence patterns and the left-right directions, considering within-subject factors. It was confirmed that there were the significant differences in the main factor (row-wise sequential pattern) for both the accuracy rate ($p < 0.001$) and response time ($p < 0.001$), but there was no significant difference in the main effect regarding the direction and the interaction of two factors.

Bonferroni's multiple comparison (comparison count: 10 times) of the accuracy rate and response time among the 5 stimulus sequences was conducted and its results were shown in **Table 1a**. The accuracy rate for the row-wise sequence from 11 to 15 was the highest and response time is shortest among the row-wise ones. This result suggests that it is easier to identify the sequence that pass the subject standing point, the loudspeaker 13 and the identification performance became higher as the distance to the sound sequence from subject decreases.

In the same way as row-wise patterns, **Figure 4c** and **d** illustrate the mean accuracy rates and response times for the column-wise patterns and the direction from front/behind, respectively. The two-way ANOVA also were conducted for the mean scores regarding the sequence patterns and the direction. It was confirmed that the significant differences regarding main effect of sequence pattern were detected in both the accuracy rate ($p < 0.001$) and response time ($p < 0.001$) for the column-wise sequence. Additionally, there were the weak significant effect regarding the direction on the accuracy rate ($p = 0.078$) and the strong significant effect on the response time ($p = 0.0068$), but the interaction was not detected. Results of Bonferroni's multiple comparison (comparison count: ten times) of the accuracy rate and response time regarding the column-wise sequences pattern is shown in **Table 1b**.

The accuracy rate for the sequence from 3 to 23 was the highest and the response time is shortest among the column-wise ones. In the same as results regarding the row-wise sequence, it is easier to identify the sequence that pass the subject standing point, the loudspeaker 13 and the identification performance became higher as the distance to the sound sequence from subject decreases. However, the performances (accuracy rate and response time) for the sequences in the second and fourth line were as well as the third centered line. These results suggest that there may be a range in which people could properly identify the location and the direction of the sound sequence.

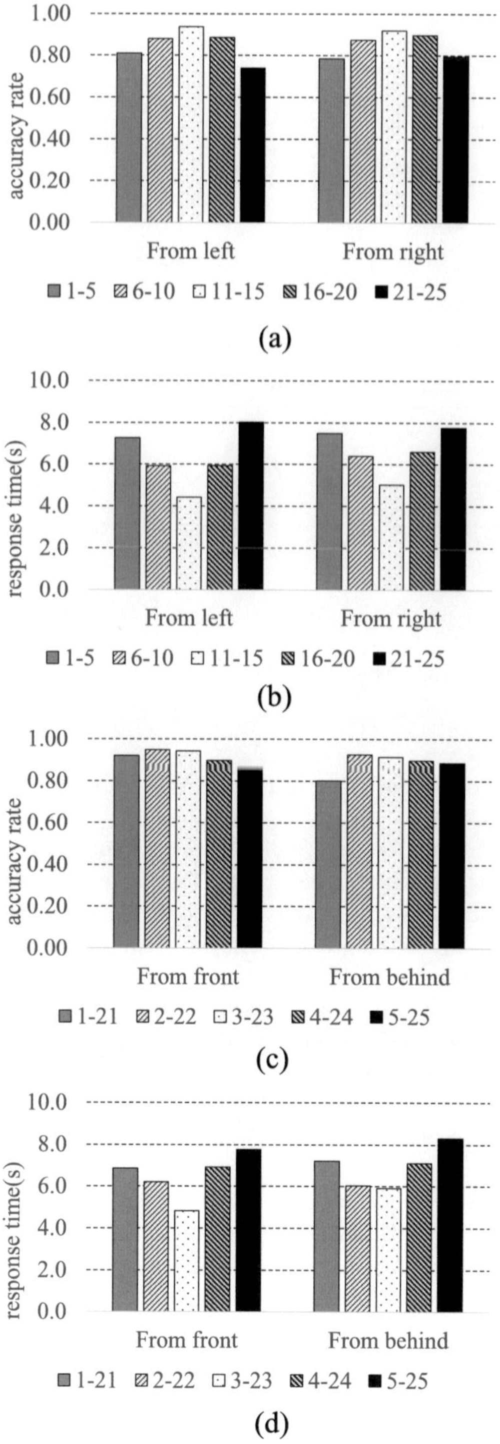

Figure 4.
Mean accuracy rates and response times for horizontal and vertical patterns in a straight line. (a) Mean accuracy rates regarding row-wise pattern from left and right. (b) Mean response times regarding row-wise pattern on the left and right hand sides. (c) Mean accuracy rates regarding column-wise pattern from left and right. (d) Mean response times regarding column-wise pattern from left and right. (From Miyoshi [6]).

Accuracy rate					Response time				
(a) p values for pair-wise comparisons in row-wise pattern									
Patterns	1–5	6–10	11–15	16–20	Patterns	1–5	6–10	11–15	16–20
1–5	—	—	—	—	1-5	—	—	—	—
6–10	0.035	—	—	—	6-10	0.002	—	—	—
11–15	<0.001	0.652	—	—	11-15	<0.001	<0.001	—	—
16–20	0.008	1.00	1.00	—	16-20	0.006	1.00	<0.001	—
21–25	1.00	0.001	<0.001	<0.001	21-25	1.00	<0.001	<0.001	<0.001
(b) p values for pair-wise comparisons in column-wise pattern									
Patterns	1-21	2-22	3-23	4-24	Patterns	1-21	2-22	3-23	4-24
1–21	—	—	—	—	1-21	—	—	—	—
2–22	0.007	—	—	—	2-22	0.076	—	—	—
3–23	<0.001	1.00	—	—	3—23	<0.001	0.275	—	—
4–24	0.97	0.716	1.00	—	4-24	1.00	0.088	<0.001	—
5—25	1.00	0.057	0.18	1.00	5–25	0.032	<0.001	<0.001	<0.001

Table 1.
Results of Bonferroni's multiple comparison for accuracy rates and response times in cases of row-wise and column-wise patterns.

3.2 Identification of sound emitting in right-angle sequences

The first experiment investigated the identification performance of the subjects for the straight sequence of the emitted sound. The actual evacuation paths to exit include the straight and right-angle paths. For example, an evacuee evacuates from the inside of a building to the outside by going straight and turning. The identification performance for the emitting sound in right-angle sequences must be evaluated to ensure that the proposed guidance system works effectively in the event of a disaster. In this section, we summarized the second experiment to evaluate the performance of identification for right-angle sequences based on a reference [6].

3.2.1 Methodology of experiment 2

In the second experiment, the sequence of sound stimuli was generated in the same experimental environment with the first experiment as shown in **Figure 1**. Twenty-five loudspeakers were arranged in a 5 × 5 grid 4 m above from floor level. For this experiment, the distance between the loudspeakers was fixed at 3 m to compare the emission patterns of the straight and right-angle lines. Considering the three experimental factors: sequence shape (two levels of straight and right-angle sequences), stimulus type (voice and swept-sound), and emission-time interval (0.5 and 1 s), the experiment was conducted to assess whether or how these factors affect identification of the sound sequence. The second experiment was conducted under eight experimental conditions, including all combinations of the above three factors. In each trial, sound stimuli were emitted in four right-angle and two straight-line sequences. In addition, the both directionalities of all the sequences, front-to-back one and back-to-front one, were considered, as illustrated by green and bidirectional arrows in **Figure 5**.

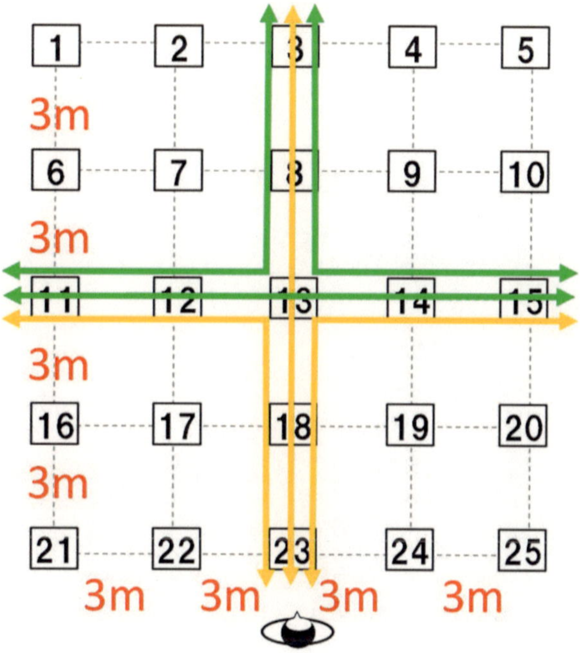

Figure 5.
Arrangement of audio loudspeakers, subject position, and sequential pattern of sound stimuli in the second experiment.

The subjects were instructed to stand at a specific position near by the loudspeaker 23 and listened to and identified the sequence pattern as quickly as possible. The sound sequences were randomly selected from 12 possible sequential patterns and start position of sound sequence was determined randomly. The sound stimulus continued until the subjects returned their answer. An experimenter recorded the participants' response time and the accuracy of their answers (accuracy rate). The subjects of the second experiment were seven male students (20–21 years old) who had participated in the first experiment.

3.2.2 Experimental results and discussion of experiment 2

Firstly, the mean accuracy rates and response times for each stimulus type and emission-time interval in the second experiment were illustrated in **Figure 6**. It was confirmed that there were the significant differences in accuracy rates for two factors: stimulus type ($p < 0.01$), emission-time interval ($p < 0.001$), and their interaction ($p < 0.01$), as the result of a two-way ANOVA considering the within-subject factors. Furthermore, there were significant differences in the response times for stimulus type ($p < 0.05$) and emission-time interval ($p < 0.05$).

Comparing the identification performance between sound types, the accuracy rate of identification for sound sequence using voice with emission-time interval 0.5 s was less than other cases, but the one for the other conditions was almost 100% because the task was performed completely. The identification of the emitting patterns of voice became difficult when the emitting voice stimulus switched to another speaker in the middle of the phrase. The performance in these emitting conditions became lower than in the other conditions.

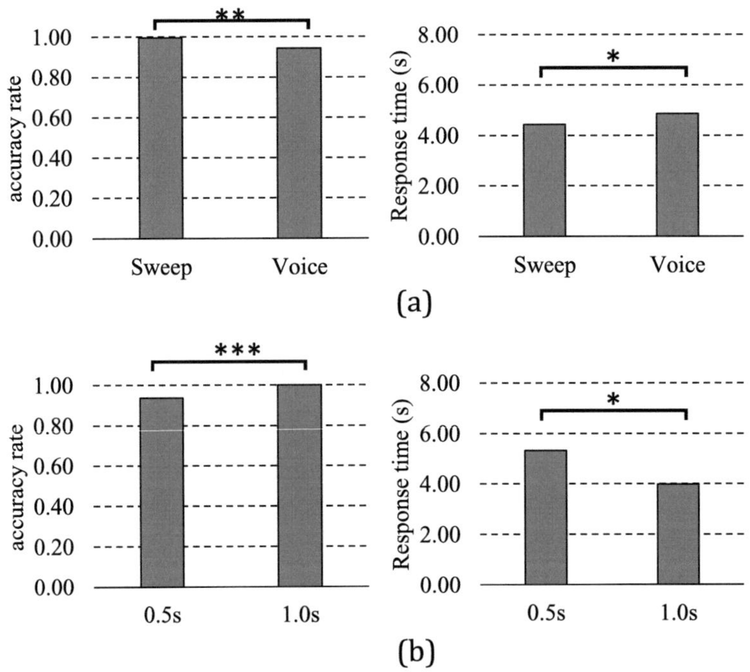

Figure 6.
Mean accuracy rates and response times regarding two experimental factors. (From Miyoshi [6]). (a) Mean accuracy rates and response times regarding stimulus type. (b) Mean accuracy rates and response times regarding the delay of stimuli.

Comparing the results for the sequences emitting along the right-angle pattern to the one emitting along the straight-pattern (**Figures 3** and 7), the mean-accuracy rates were higher for the emitting pattern at the right angle than in the straight line. In the second experiment, the subjects stood nearby the loudspeaker 23 and identified the sound sequences emitting in front of them. This is because identification performance for the sequence emitting in front of subjects was better than the backward ones as evaluated in the first experiment.

Next, the results regarding the sequence patterns are described. The mean accuracy rates and the response times, classed according to the sequence pattern and the spatial proximity to the stimulus from subject standing position were illustrated in **Figure 7**. The 12 sound sequences were classed into two level in the spatial proximity, "near" and "far" that were indicated with yellow and green lines in **Figure 5** respectively.

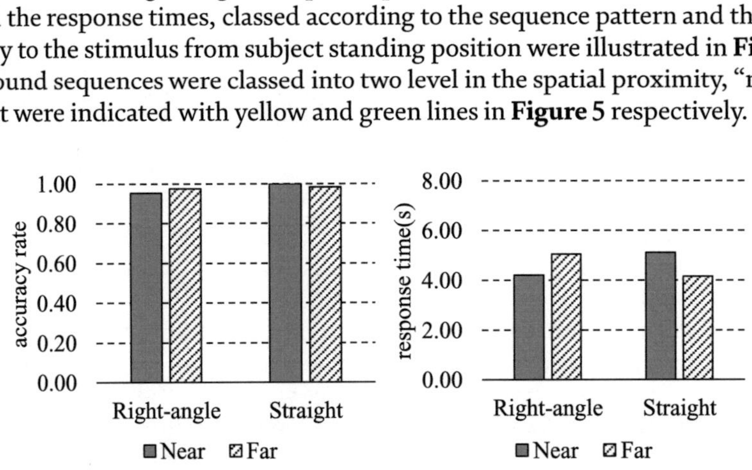

Figure 7.
Mean accuracy rates and response times regarding the spatial proximity to sequence. (From Miyoshi [6]).

A two-way ANOVA was used to estimate how the means of the accuracy rates and the response times changed according to the levels of two factors, the sequence patterns and spatial proximity of stimuli from subjects. It was not confirmed that there was neither significant difference in the main factors nor in their interactions. The accuracy rate was almost 100% in all the cases. These results mean that there was no difference in the accuracy rates and response times depending on the sequence pattern, and that the subjects could perceive the sequences of acoustic stimuli emitted in front of them and identify the sequence pattern almost completely in cases where the emission pattern is a straight or right-angle. Therefore, the subjects might be able to follow the acoustic sequence of the emitting stimulus to the emergency exit more quickly even if the evacuation path is more complicate route including straight and right-angle paths. In discussing the practicality of evacuation guidance systems, the results are preferable for realizing a guidance system that emits acoustic stimuli along a predetermined evacuation path.

However, we observed no difference between the sequence shapes, which could be because there was a consistent distance between the loudspeakers. During an actual emergency, the evacuation path may contain a series of short paths to the exit. Therefore, in future, we will verify the subjects' performance with a smaller distance between the loudspeakers.

4. Ability to guide evacuees by emitting sound along predetermined paths

4.1 Experimental condition of experiment 3

We conducted the third experiment to investigate whether the subjects were able to follow the sound emitted along the evacuation paths. Additionally, we investigated whether it is possible for people to follow complex routes that have a lot of turning points on it. In this experiment, the complexity of the guidance is defined by the number of turning points on itself. As the number of turning points included in a guidance patten increases, it gets more complicated. In case of moving along the more complex path on which subjects turn to the right and left repeatedly in a short time, more accurate and quick sound localization is required to identify the sound stream. Therefore, we defined the complexity of the evacuation path by the number of turning points on it as the difficulty to identify and follow the acoustic stimuli.

In this experiment, the configurations about the locations of loudspeakers and sound sources were the same as those in the previous experiments. **Figure 8** shows the experimental configuration and subjects following the sequence. The distance between the loudspeakers and their heights were set to 3 and 4 m, respectively.

The three factors considered important to the performance of the experimental task were the type of sound source, the type of loudspeaker, and the patterns of emitting sound sequences in the third experiment. Two levels, voice and swept-sound, were set with respect to the type of sound source. In the same way with previous experiments, the phrase in female voice, "Here is an emergency exit" was used as voice sound and the acoustic stimuli whose frequency changed from 500 to 1000 Hz continuously was used as the swept-sound.

The two levels, a capacitor-type speaker and a dynamic range one, were set with respect to the type of loudspeaker on which the sound stimuli were emitted. The capacitor-type flat and the conventional dynamic rage speakers have different specifications for directionality, which is the property to focus audio and deliver clear sound

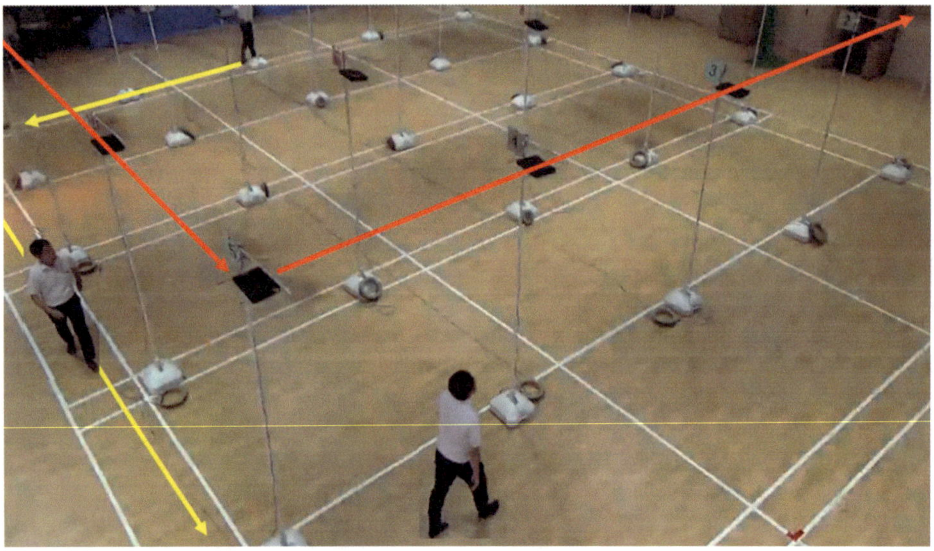

Figure 8.
Experimental configuration and a subject following the sound sequence.

precisely where it is needed. In other words, the reduction in sound level through the capacitor-type speaker is smaller even when the reach is farther away because sound spread is smaller than the dynamic one. The capacitor-type one has higher performance than the dynamic range one in the specification of directionality. This factor was designed to evaluate whether the directionality of the loudspeaker affects the sound localization performance for a moving sound source.

The third factor in this experiment is the pattern of emitting sequences. **Figure 9** shows the emitting sequence patterns of sound, which are drawn as connections of consecutive column-wise and row-wise line segments. "S" and "G" in **Figure 9** indicate the start and goal points of each sound sequence. The sequence patterns are classified into five categories based on the number of turning points in themselves. There were 20 sequences (two sequences in conditions of the number of turning points and start-goal places), as shown in Figure 9. A sequence with one turning point means that the subjects turn for direction once during following it. The third experiment was finally designed under all the combination conditions of three factors (2 levels × 2 levels × 20 patterns) as described above.

4.2 Experimental procedure of experiment 3

The sound source level of voice and swept-sound were set to 80 dB (A-weighted loudness level) at 1 m from the loudspeaker. The noise level was measured using an integrating average-type sound-level meter (LA-1441, Ono Sokki). The sequences of sound were emitted from the lower left (loudspeaker 21) or lower right (loudspeaker 25) to the upper right (loudspeaker 5) or upper left (loudspeaker 1) which were indicated by circle "G" are shown in **Figure 9**. The evacuation routes with one to five turning points are also illustrated by the solid and dotted lines in **Figure 9**, which were formed symmetrically with respect to the diagonal.

The subjects were seven students (six males and one female, 20–21 years old) with no hearing abnormalities during their annual medical examination. The subjects were instructed to follow the sound sequence at walking speed. The sequences of emitting

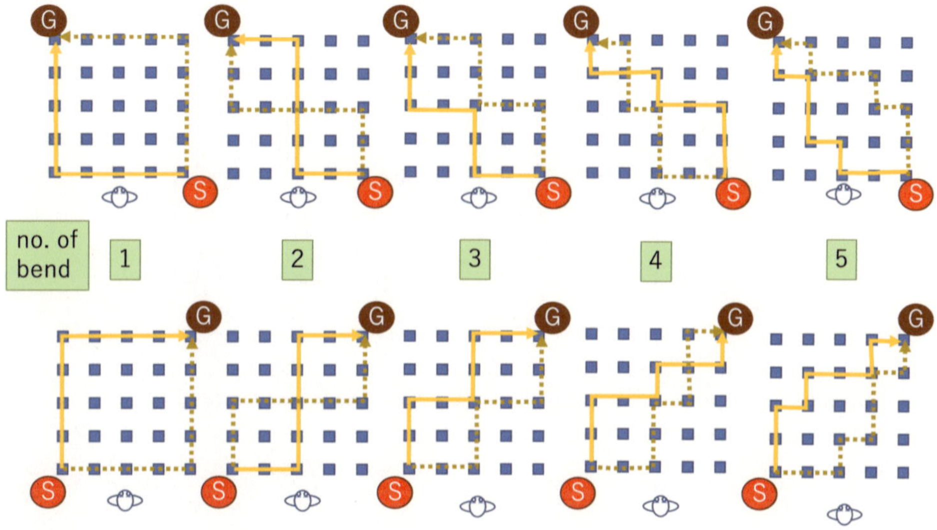

Figure 9.
Sound-spatial sequences provided to subjects.

sound were randomly presented to them. In each trial the first point of the sequence was randomly chosen so as not to infer the emitting sequence pattern based on it. The subject stood up at the point under the loudspeaker 23, and identified the sequence of emitting sound and followed it.

The experiment using voice as the sound source was conducted first, and one using swept-sound was conducted two months later. In each experiment the subject performed the trial to identify and follow the sequence of sound that was emitting on two types of loudspeakers, the capacitor-type and the conventional dynamic range speakers. The half of subjects (five subjects) performed the trial using the capacitor-type speakers at first and the one using the conventional dynamic range one after it. The remaining subjects (three subjects) performed the trial in reverse order with respect to the type of loudspeakers.

4.3 Experimental results of experiment 3

4.3.1 Results of the ability to follow the emitting sound

In the third experiment, we recorded the success or failure of following the sequences of the emitting sound and time required to follow the sound from the start and end points. The following three experimental factors were considered in this experiment: the type of sound source (voice and swept-sound), type of loudspeaker (capacitor-flat and dynamic ones), and complexity of the sequence that was defined from one to five by the number of turning points.

The mean values of the time required to follow a sequence and the success rate for each factor are shown in the **Figures 10–12**. A three-way ANOVA within-subject was used to estimate how the means of the success rates and the required times changed according to the levels of three factors as mentioned above. No significant differences were observed in the success rates for all main factors and their interaction. In the result with respect to the required time, there was no significant difference in the type of sound source (p = 0.457), but there was a significant difference in the type of

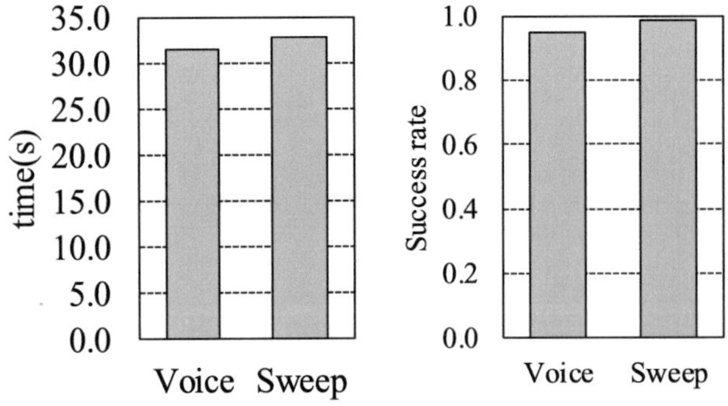

Figure 10.
Mean values of the time required to follow a sequence and the success rate for sound sources.

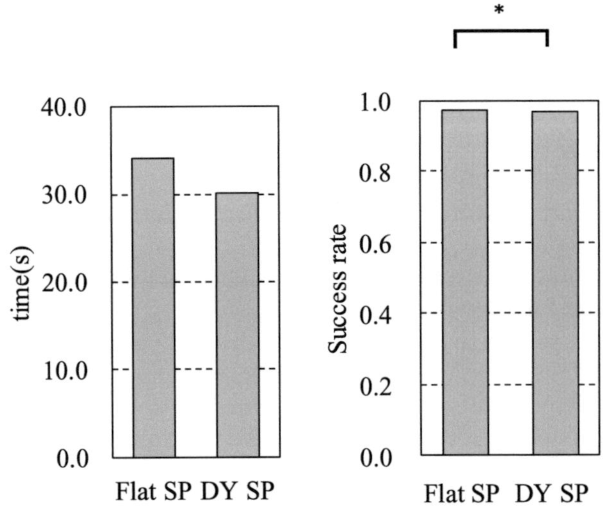

Figure 11.
Mean values of the time required to follow a sequence and the success rate for loudspeaker types.

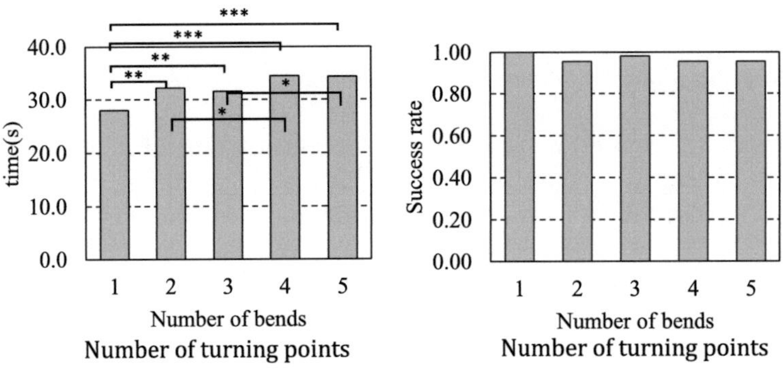

Figure 12.
Five-level Bonferroni's multiple comparison of the complexity of the emitting patterns.

loudspeaker ($p = 0.0415$) and the complexity of the emitting pattern ($p = 0.0013$). **Figure 12** shows that, a five-level Bonferroni's multiple comparison of the complexity of the emitting pattern showed a significant difference.

The average of success rate was 0.970 for all experimental condition, which shows that people were able to identify and follow the sequence of sound in almost all cases. Therefore, there were no significant differences among the sound source, loudspeaker type, and sequence patterns. Even though the success rate of the following is high, the required time is considered as the measure for the difficulty to identify and follow the emitting sound source. In the third experiment, we compared the time required to follow the emitting sound source because the success rate of following the sequences of sound sources was extremely high and no significant difference was observed between the factors.

4.3.2 Experimental results with respect to sound sources

Figure 10 shows that there was the difference in the required following time between types of sound source and it is slightly longer for swept-sounds than voice. However, the analysis of variance showed no significant difference between them in contrast to the results of the first experiment. The task in experimental trial is identifying and following the sequence of emitting sound. The subject is required to repeatedly perform the sound localization for moving sound on the loudspeaker near own current location in the task. The experimental results that the success rates were extremely high illustrate that the subjects were able to perform the continuous sound localization precisely even though the complexity of the guiding routes were relatively high. Therefore, this result suggests that the proposed guidance system using the emitting sound is effective and feasible to lead the predetermined evacuating route even if the voice was used instead of the swept-sound as the acoustic source.

4.3.3 Experimental results with respect to loudspeakers

Figure 11 shows the mean of the required time for following the sequence of the emitting sound on two type loudspeakers, the capacitor flat one and the conventional dynamic range one. The required time is shorter in using the dynamic speaker than the capacitor flat speaker. This result suggests that subjects are able to perform sound localization for a sequence of the sounds emitting on the dynamic range speakers more easily than on capacitor flat one, also were able to follow them. Some subjects commented that they could listen to sounds on the dynamic speaker more clearly than the capacitor one. This tendency is obvious for the sounds emitting on speaker far from their current position. In the task of the third experiment subjects were required to repeatedly perform sound localization for the sound source in spatially wide area. The capacitor flat speaker has stronger directionality than the dynamic range one, then the sound emitting on the dynamic range speaker acoustic stimuli spread more widely and was easier to catch up than on the capacitor one. It is reasonable to assume that the difference of the acoustic property between two types of speakers results the difference of the required time to follow.

4.3.4 Experimental results of sequence patterns

Figure 12 shows a comparison of the required time and the success rates according to the complexity of sequence patterns. There is a little difference in success rates but significant difference ($p = 0.0013$) in the time required to follow the sequences among the complexity levels by the ANOVA considering the within-subject factors. It was confirmed that the required time increased as the number of turning points increased, i.e., as the complexity of the emitting pattern became higher.

The Bonferroni's multiple comparison (comparison count: 10 times) were conducted for the five complex level of sequences after confirming the significant difference. According to the result, there was no significant difference in the required time between the cases of three and four turning points and between four and five turning points. This result suggests that the difficulty of following the pattern increases as the number of turning points increases.

As the number of turning points in the emitting pattern increases, the success rate slightly decreases, and the required time increases as shown in Figure 12. This result suggests that the evacuation performance decreases as the pattern becomes more complicated in the proposed evacuation system.

5. Conclusion and future works

In this study, the subjects' ability to identify the location and direction of acoustic spatial sequences and follow it was evaluated through three experiments to discuss the practicality and feasibility of the proposed evacuation guidance systems.

In the first and second experiments, the accuracy rate and response time of subjects for identification the different sequences of sound stimuli were compared among several experimental conditions combining factors: the stimulus the type, emission interval, the distance between loudspeakers, and the sequence patterns. In the first experiment, the accuracy rates improved when the voice stimulus was used and when the emission-time interval was extended. Additionally, it was confirmed that the identification performance becomes better as the distance from the position of subject gets shorter.

In the second experiment, we observed no significant difference in the accuracy rates and response times for different sequence patterns (straight line and right-angle) under the experimental conditions.

In the third experiment the ability of people to follow the sequences of the emitting sound was evaluated based on the success rate and the required time to do so. The three experimental factors were considered, which were the stimulus type, the type of loudspeaker, and complexity of the sequence measured by number of turning points on it. In the third experiment, people took more time but could follow the sequences of emitting sound, which included five turning points. The required time is shorter in using the dynamic speaker than the capacitor flat speaker. This result suggests that subjects can perform sound localization for a sequence of the sounds emitting on dynamic speakers more easily than the capacitor flat one.

The results of this study demonstrate the practicality of an evacuation guidance system using a sound sequence that emits specific sounds on a set of loudspeakers. The factors affecting the performance of subject's identification of the acoustic stimuli were examined and analyzed, but the level of factors dealt with was limited such as the frequency change region of the swept-sound. Therefore, a more detailed analysis of the degree of influence of each factor is needed in the practical application of the proposed guidance stem for further studies.

Acknowledgements

This research was funded by JSPS Grant-in-Aid for Scientific Research (C) 19K04937. I express my gratitude here.

Author details

Tetsuya Miyoshi
Hannan University, Matsubara, Osaka, Japan

*Address all correspondence to: miyoshi@hannan-u.ac.jp

References

[1] Jin T. History of the emergency exit lighting sign [in Japanese]. JLMA Journal, Japan Lighting Manufacturers Association. 2015;**13**:17-23

[2] JLMA. Japan Lighting Manufactures Association, 1993 JIL5505 Technical Regulation of Evacuation Guidance Systems [in Japanese], JLMA

[3] Yamasaki Y et al. Basic investigation of emergency guide by sound [in Japanese]. Reports of the meeting of the Acoustical Society of Japan. 1990. pp. 659-660

[4] Ito Y, Ohgushi K. An intensive system for escape guidance using voice. Science and Technology in Lighting. 1995;**79**(9):532-537

[5] Fujikawa T, Oshigami Y, Maeda N, Tomita M, Aoki S. A study on the precedence effect using evacuation signal. Acoustical Science and Technology. 2013;**34**(5):348-350

[6] Miyoshi T. Evaluation of recognition performance of guidance direction presenting by sequential emitting sound in evacuation guidance system. In: INTER-NOISE and NOISE-CON Congress and Conference Proceedings, InterNoise21, 1-5 August 2021. Vol. 263, No. 1. Washington, D.C.: Institute of Noise Control Engineering; 2021. pp. 5335-5345. DOI: 10.3397/IN-2021-3052

[7] Buribayeva G, Miyachi T, Yeshmukhametov A, Mikami Y. An autonomous emergency warning system based on cloud servers and SNS. In: Procedia Computer Science Proceedings of the 19th International Conference on Knowledge Based and Intelligent Information and Engineering Systems, 7-9 September 2015. Vol. 60. Singapore: Elsevier; 2015. pp. 722-729. DOI: 10.1016/j.procs.2015.08.225

[8] Lujak M, Billhardt H, Dunkel J, Fernández A, Hermoso R, Ossowski S. A distributed architecture for real-time evacuation guidance in large smart buildings. Computer Science and Information Systems. 2017;**14**(1):257-282. DOI: 10.2298/CSIS161014002L

[9] Yamasaki Y et al. Development of prototype system using iOS terminals for Emergency Rescue Evacuation Support System (ERESS). In: Proceedings of the 45th international conference on Parallel processing Workshops, IEEE. 2016. DOI: 10.1109/ICPPW.2016.17

[10] Hara T, Kasabe M, Kasahara M. On path selection considering length and reliability for automatic evacuation guiding based on interactions between evacuees and their mobile nodes. IEICE Technical Report. 2017;**116**(484):517-522

[11] Gardner MB. Historical background of the Haas and/or precedence effect. Journal of the Acoustical Society of America. 1968;**43**(6):1243-1248

[12] Fujikawa T, Aoki S. An escape guiding system utilizing the precedence effect for evacuation signal. Journal of the Acoustical Society of America. 2013;**133**(5):3362-3362

[13] Aoki S, Maeda N, Fujikawa T. Evacuation system utilizing the precedence effect. Applied Acoustics. 2020;**163**(June). DOI: 10.1016/j.apacoust.2020.107233

Chapter 6

Noise Profile Categorization for Noise Mapping in Cities: The Case of Cuenca (Spain)

José A. Ballesteros, María Jesús Ballesteros, Samuel Quintana and Marcos D. Fernandez

Abstract

According to the European Law, noise maps in cities have to be worked out and updated every 5 years. Because of this, it is interesting to establish new methodologies to develop and update the noise maps in a more efficient way. Although there are specific standards to carry out noise maps and a good practice guide was defined, there is not a common procedure in the definition of the noise map. In each research, a specific methodology is defined based on the experience of the researchers and the characteristics of the town. In this work, a methodology based on a street typology classification is proposed to be applied to noise maps. This methodology allows allocation of the mean power and the temporal behavior to each street from its characteristics and the time profiles measured with semi-permanent noise monitoring systems. The methodology was developed, tested, and validated in the city of Cuenca (Spain) and the results obtained are shown in this chapter.

Keywords: urban noise, noise mapping, noise profiles, street typologies, monitoring system

1. Introduction

The European Law [1], developed in Spain by the Royal Decrees [2, 3], states that noise maps in cities with a population in excess of 100,000 individuals need to be developed and updated every 5 years.

A noise map can be done just based on a prediction model, that is, without any kind of empirical measurement; but to achieve a minimum accuracy level, a calibration of the model is required through a selected number of in-situ measurement points, which increases the budget to carry out the noise map. Moreover, although specific standards and a good practice guide [4] have been defined to be used in noise maps, each research group defines its own methodology based on its experience and the characteristics of the town. Therefore, a proper balance must be found among the cost of the noise map, its accuracy, and the number of measurement points; that balance can be improved with the use of a road categorization system to achieve representative measurements of all the types of roads to minimize the cost without losing accuracy in the model.

Classically, four sampling methods have been considered: to select the sampling points by laying a grid over a map of the target zone, to define a source-oriented sampling where measurement locations are selected arbitrarily to represent different road and traffic conditions, to define a receptor-oriented sampling in which the noise exposure of a particular class of receptors is investigated, and to select the sampling points using a prior classification of the urban noise [5].

In [6] sampling points were selected, covering all of the representative locations and all of the streets in the area. The sampling points were always in the middle of the street or location-specific. This methodology for selecting sampling points and locations allows performing a detailed study of the area and it is not based on a previous categorization of the streets, but neither is similar to the commonly used grid method.

Preliminary studies of three towns [7–9] served to adapt and optimize the definition of the categories established in [10]. This definition consisted of six categories:

- Class 1: national roads
- Class 2: streets that provide access to the major distribution nodes of the town
- Class 3: streets that lead to regional roads
- Class 4: streets that allow clear communication among the previous types of streets
- Class 5: the rest of the streets of the town except walking streets
- Class 6: walking streets

The categorization method was later applied in [11] to five medium-sized Spanish towns with populations ranging from 218,000 down to 50,000, and later in [12] to 20 towns with sizes ranging from 2200 to 700,000 inhabitants, and with areas between 0.57 and 59 km^2.

In this work, a methodology based on a street typology classification is proposed to carry out noise maps. This methodology, recently validated in [13, 14] allows allocating the mean power and the temporal behavior to each street from its characteristics and the time profiles measured with semi-permanent noise monitoring systems. Because of this, as the streets are classified a priori in different typologies, the number of measurement points is reduced maintaining the accuracy of the noise map and reducing the measurement cost. The proposed methodology has been applied to the noise map of Cuenca.

2. Methodology

2.1 The town of Cuenca

Cuenca is a Spanish town in the Castilla-La Mancha region, capital of the province of the same name. The town's mean altitude is 946 m above sea level (a.s.l.). The area of the municipality is 911.06 km^2, and the total population of the city is 54,898 inhabitants (year 2018) [15].

The town has two different areas: the old town and the new one. The first one is located in the high part of the town, bordered by rivers Júcar in the north and Huécar in the south. The new part of the town is located at the west-south of the old town towards north-south.

Cuenca has an important historical and architectural heritage and many museums in the old town, been considered World Heritage City since 1996.

2.2 Road type definition

To define the street typologies in the town of Cuenca, the following road and urban characteristics (all of them with influence from the acoustic point of view) were taken into account:

- Road characteristics:
    - Number of lanes of the street.
    - Average daily traffic (ADT)
    - Average daily speed (ADS)
- Urban characteristics:
    - Road width
    - Pavements
    - Main use of the street.

Based on the aforementioned characteristics, seven street typologies were established in this research:

- Type 1 (class 1): Motorway.
- Type 2 (class 2): Main distribution streets in the town.
- Type 3 (class 4): Secondary distribution streets in the town.
- Type 4 (class 3): Main distribution streets in a neighborhood. These streets are used for traffic distribution in a neighborhood and are connected with bigger streets.
- Type 5 (class 4): Residential distribution streets. These streets are used by people to arrive in a completely residential street (types 6 and 7).
- Type 6 (class 5): Residential streets in the town and residential distribution streets in new neighborhoods.

- Type 7 (class 5): Completely residential streets. Only the people who live in these areas use these streets.

This street classification allows performing a categorization of the streets in a town based on their acoustics characteristics. Thus, measurements only in some streets of each category are needed, being possible to apply the noise profile obtained for each category to the rest of the streets in the same category during the noise map simulation phase. According to [16] these procedures only generate little deviations from strictly using only real measurement data for all the sampling points.

2.3 Long time measurements

To carry out long measurements, three noise monitoring equipment have been placed in fixed locations during 11 months, and another one has been moved among different locations. This hybrid procedure is quite common to increase the temporal and spatial resolution [17].

The location of two of these fixed equipments has been chosen to take into account the main areas of the town: Old town and Carretería Street. The third has been located to characterize the residential noise of the town (Acacia Uceta Street).

The moving equipment has been located in different places in order to characterize the noise in different kinds of streets in the town. The locations of this equipment were: República Argentina Avenue, 16 (3 months), Mediterráneo Avenue (4 months), Hermanos Valdés Street, 4 (2 months).

The location of each monitoring equipment can be observed in **Figure 1**, and their accurate location is in **Table 1**.

Figure 1.
Monitoring equipment locations.

Equipment label	X coord. [m]	Y coord. [m]
Old Town	574,163.64	4,436,764.68
Carretería Street	573,634.63	4,436,174.01
República Argentina Avenue	573,414.85	4,436,005.10
Acacia Uceta Street	572,681.20	4,436,180.54
Mediterráneo Avenue	574,127.48	4,435,221.37
Hermanos Valdés Street	573,619.63	4,436,365.07

Table 1.
Monitoring equipment locations.

The equipment has been installed at 4 m height and 2 m from the facade according to the regulations [18]. For that, a bracket fixed to the facade has been used with an outdoor microphone. In **Figure 2** the microphone installation can be observed.

The microphone calibration was checked manually when the equipment was installed when it was removed, and at least once a week. If there was an error, the equipment was repaired or replaced. Moreover, four automatic CIC calibrations per day were configured at 00, 10, 14, and 19 h, which were permanently monitored and checked.

The main acoustic parameters that have been measured were:

- L_{Aeq}: A-weighed equivalent level.
- L_d: Day equivalent level.

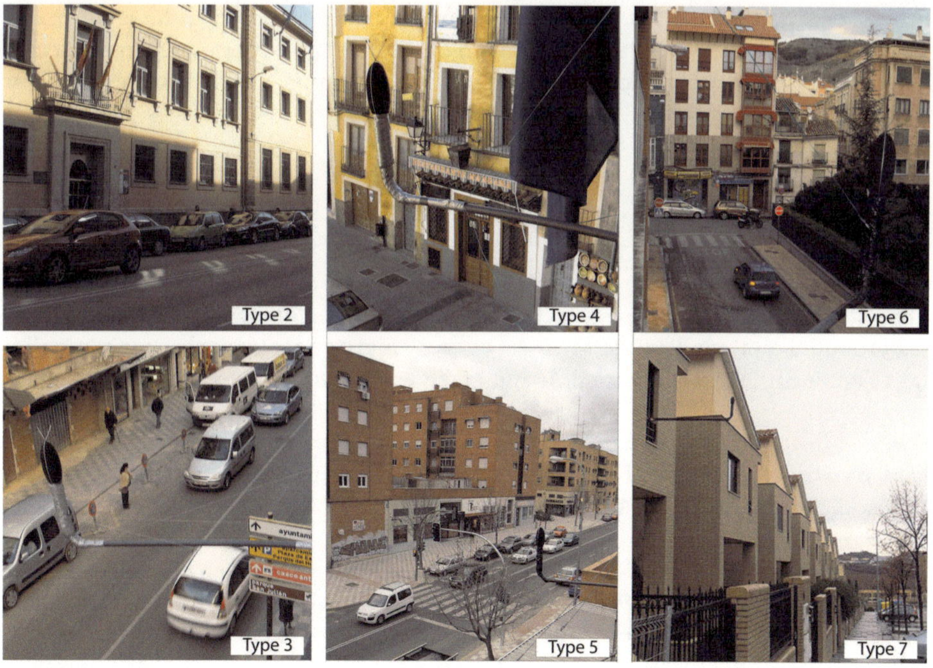

Figure 2.
Microphone installation.

- L_e: Evening equivalent level.
- L_n: Night equivalent level.
- L_{den}: Day-evening-night equivalent level.
- L_N: Percentile levels.

After taking the measurements for a whole month, a validation process was carried out. First of all the automatic calibrations were checked. If there was any error, data between the last correct calibration and the first one (the wrong calibration would be in the middle) were removed because it is not possible to know when the equipment failed. For instance, if calibrations at 10 h and at 19 h were right, and calibration at 14 h was wrong, data from 10 to 19 h were removed.

Then, the meteorological conditions were checked [18]. If it was raining, data for this period of time were removed. Moreover, if the average speed of the wind during a certain day was higher than 4 m/s (it is more restrictive than the regulations −5 m/s-), data for this day were also removed.

After these validations, a more detailed analysis of the L_{Aeq} parameter was done. Analyzing the time profile and the audio files, anomalous events have also been removed, for instance, if events in the street that are not representative of the typical behavior of the area took place (parades, parties, etc.).

With this, all the data considered for the studio could be assessed as correct.

Once the data have been validated, three different hour profiles have been obtained each month, the whole month profile, the working day profile, and the non-working day profile. To obtain these profiles, data for each hour have been averaged. The profiles obtained for all the months during the period of time when the equipment was in a certain location were similar. With this, all the hour profiles were also averaged to obtain the typical hour profile of the area.

2.4 Road simulation process

The aim of the noise map simulation is to make up a digital model of the town to obtain the noise propagation levels due to the traffic. The digital model is composed of: a ground digital model with the level curves (**Figure 3a**); obstacles, mainly buildings (**Figure 3b**); roads (**Figure 3c**), and receivers (**Figure 3d**) to evaluate the noise parameters.

When the digital model has been carried out, the next step consists of defining the characteristics of each one of the layers stated. Concerning roads, the data to define in the model are the ADT and ADS for light and heavy vehicles during the 3 day periods (day, evening, night).

To make this process easier, all the roads in the town have been classified according to the typologies defined in Section 2.2. The map in **Figure 4** shows the classification carried out. Then, the simulation results are obtained and compared with those from long time measurements and short-time measurements, and the noise map is adjusted.

DOI: http://dx.doi.org/10.5772/intechopen.104396

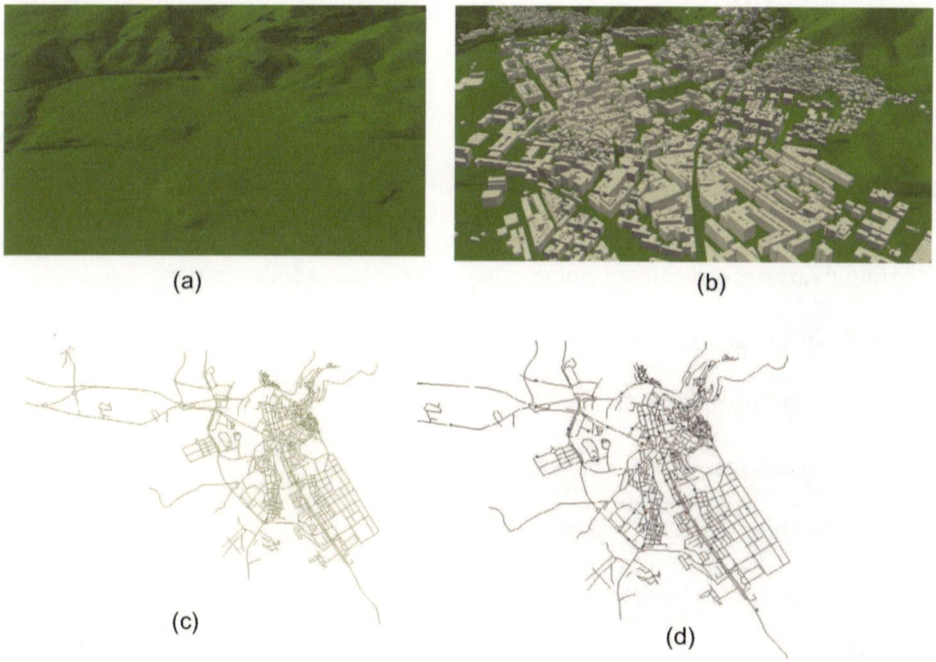

Figure 3.
Noise map process. (a) Ground layer. (b) Building layer. (c) Road layer. (d) Receivers.

Figure 4.
Road classification.

3. Results

In this section, results obtained for the measurements carried out with the noise monitoring systems in each one of the chosen locations are shown, together with the discussion of the whole results.

The street typologies defined in Section 2.2 have been checked out with short time measurements before the installation of the monitoring systems in the proposed locations (type 1 roads are not set, as there were no people living near a motorway). Accordingly, the measurement points can be classified as follows:

- República Argentina Avenue: Type 2.
- Carretería Street: Type 3.
- Old Town: Type 4.
- Mediterráneo Avenue: Type 5.
- Hermanos Valdés Street: Type 6.
- Acacia Uceta Street: Type 7.

As there are very few differences among the measurements for every month, only the averages for the whole measurement time are shown. **Figures 5** and **6** illustrate, respectively, the weighed equivalent levels for day and night periods (results for the evening period are almost identical to those of the day period). **Figure 7** shows the mean L_{Aeq} level in dB(A) for each one of the measurements points. **Figures 8** and **9** show the mean L_{Aeq} levels during working days (Monday to Friday) and non-working days (weekends and festivities), respectively. Finally, **Figure 10** represents the noise climate (L_{10}–L_{90}) and **Figure 11** the singular noise events (L_1).

According to all these results, general and specific statements can be done that prove and justify the approach of using road typologies to simplify noise maps.

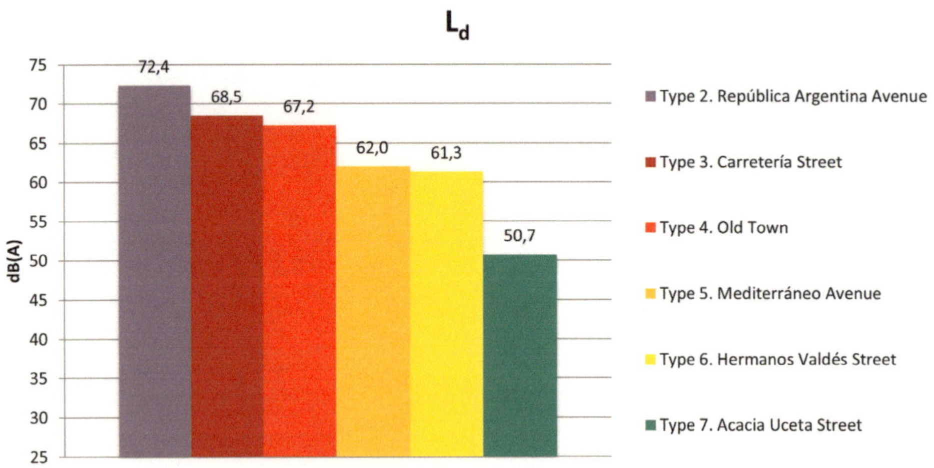

Figure 5.
L_{day} results.

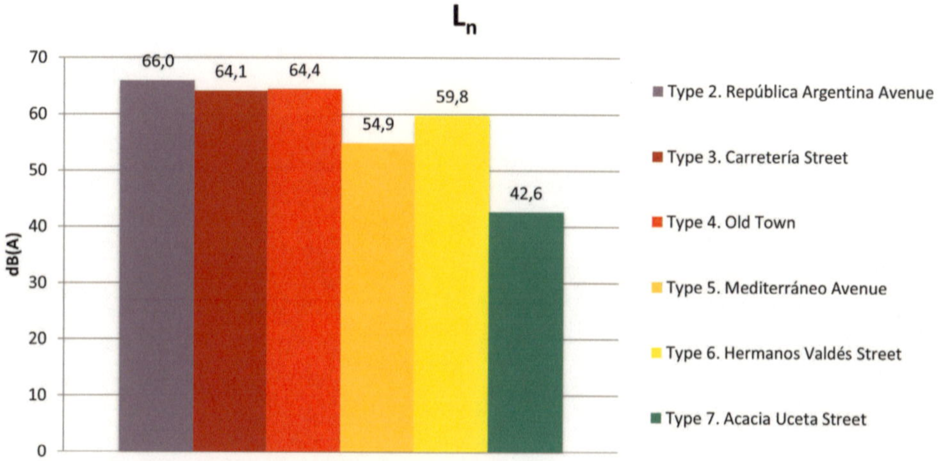

Figure 6.
L_{night} results.

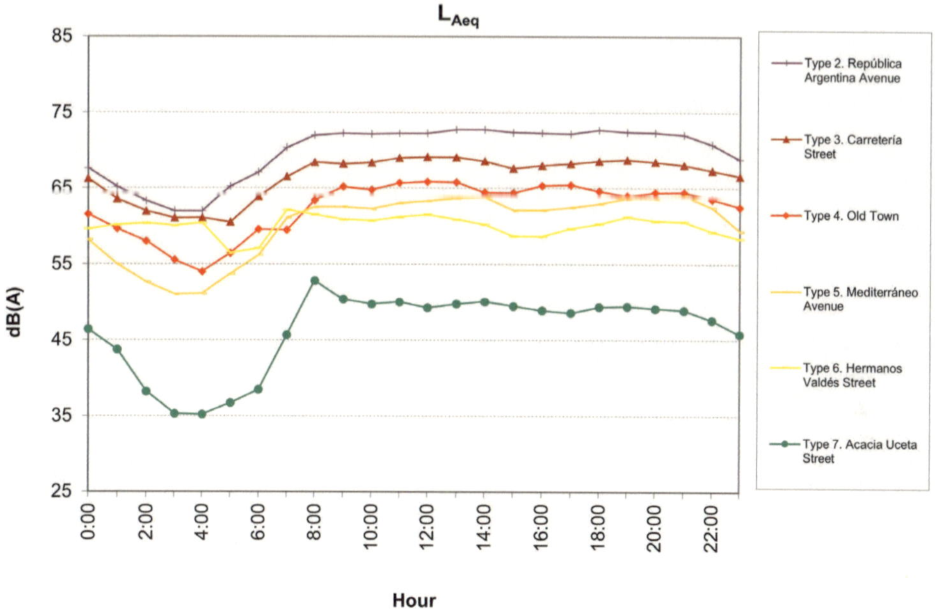

Figure 7.
L_{Aeq} results.

Firstly, specific noise profiles can be derived for each road typology, that can be described as follows:

- In roads of type 2, the L_{Aeq} is almost constant around 72 dB(A) during the day. At 22 h this level starts to decrease, being minimum from 3 to 4 h with 62 dB(A). At this time the noise level starts to increase up to 8 h. Observing the L_{Aeq} during working days, a similar trend with levels around 73 dB(A) during the day and a minimum of 60 dB(A) at 3 h can be found out. On non-working days, the level is around 72 dB(A) during the day and the minimum of 64 dB(A) is reached at 4 h.

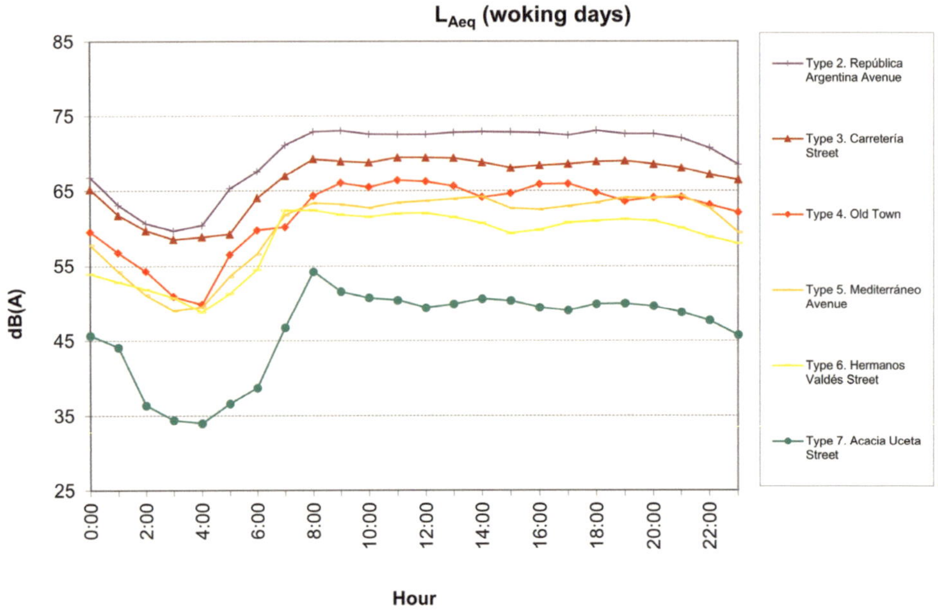

Figure 8.
L_{Aeq} results during working days.

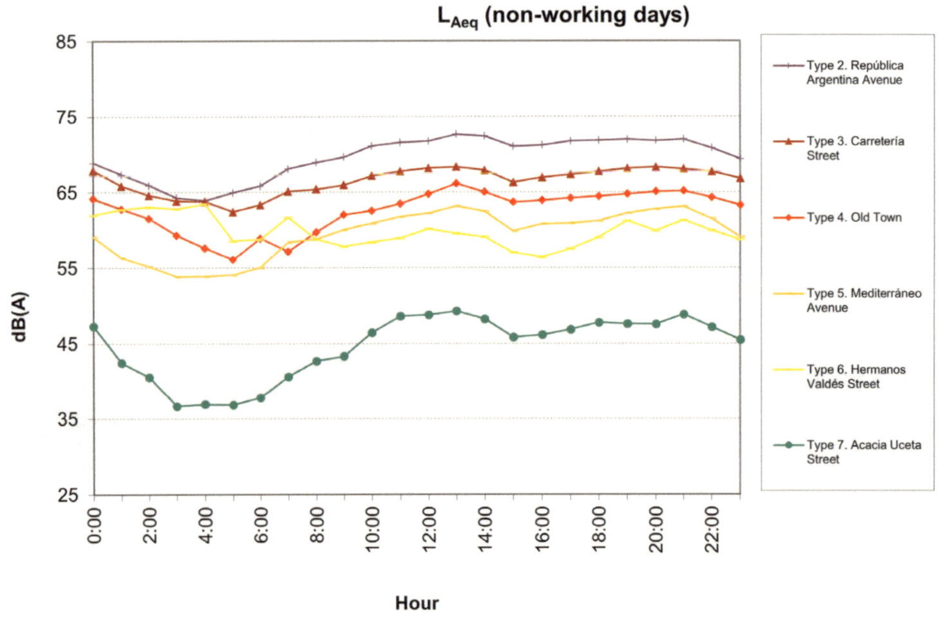

Figure 9.
L_{Aeq} results during non-working days.

- In type 3 roads, the L_{Aeq} is again almost constant close to 70 dB(A) during the day. At 22 h the level starts to go down with the minimum L_{Aeq} (60 dB(A)) at 5 h when the level starts to increase. During working days, the same trend can be

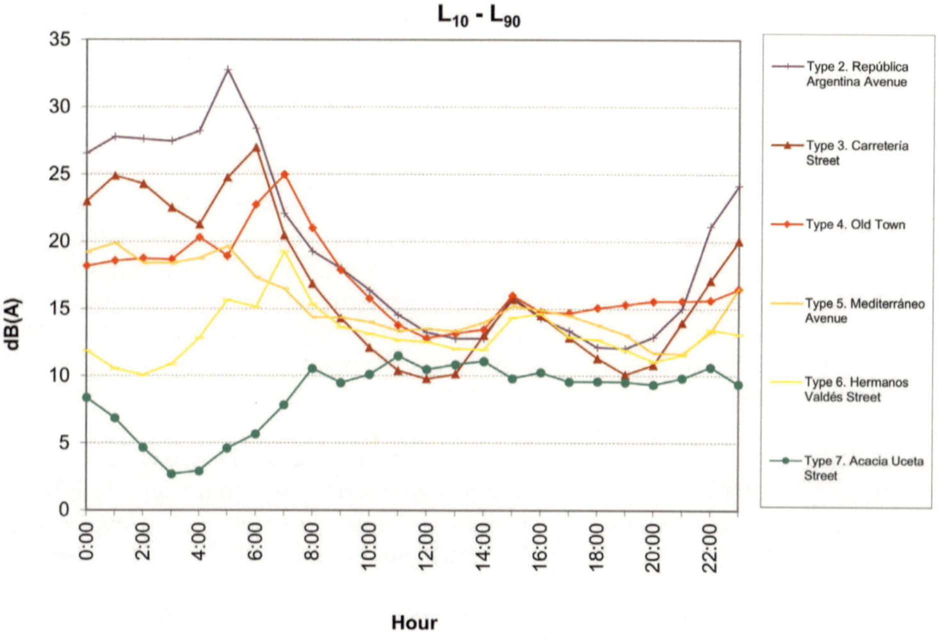

Figure 10.
Noise climate (L_{10}–L_{90}) results.

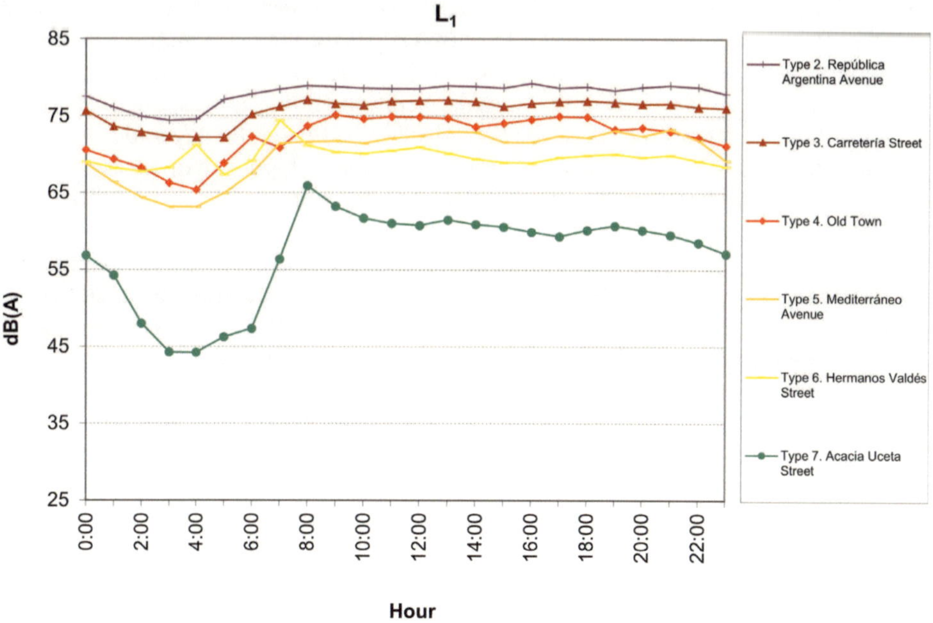

Figure 11.
Singular noise events (L_1) results.

observed, with levels around 70 dB(A) during the day and a minimum of 58 dB (A) from 3 to 4 h. Concerning non-working days, a more constant profile is obtained, with levels around 67 dB(A) during the day and a minimum of 62 dB (A) at 5 h.

- In roads of type 4, the L_{Aeq} is around 65 dB(A) from 9 to 21 h. At this time the level decreases to 54 dB(A) at 4 h, increasing from this time to 9 h. On working days, the day levels are around 66 dB(A) from 9 to 17 h, being 50 dB(A) the minimum L_{Aeq} at 4 h. Non-working days have an L_{Aeq} around 64 dB(A) during the day and a minimum of 56 dB(A) at 5 h.

- Type 5 roads show a L_{Aeq} between 62 and 64 dB(A) from 8 to 22 h with a minimum during this period from 15 to 16 h. From 22 to 4 h the level diminishes to 51 dB(A), moment when it starts to go up to 8 h. Concerning working days, similar levels are observed from 8 to 22 h, being the minimum of 49 dB(A) at 3 h. Levels during non-working days are lower during the day, with a L_{Aeq} from 60 to 63 dB(A) from 9 to 22 h, and higher during the night with a minimum of 54 dB (A) from 3 to 5 h.

- The level in type 6 roads is almost constant during the whole day with a L_{Aeq} around 61 dB(A). The level decrement is observed from 5 to 6 h with 57 dB(A). Working days have more differences between day and night with a level around 62 dB(A) in the morning (from 7 to 13 h) and around 60 dB(A) in the evening (from 13 to 21 h). At night the L_{Aeq} decreases to 49 dB(A) at 4 h. Non-working days have higher levels during the night, around 63 dB(A) than during the day, with levels from 58 to 62 dB(A). This behavior is due to the leisure activities during the night on non-working days.

- The level during the day in the 7th type of road is almost constant around 50 dB (A). The maximum level is at 8 h (53 dB(A)) because it is the moment when people go out to work. At night the levels decrease down to 35 dB(A) from 3 to 4 h. Working days have a similar profile, being the minimum L_{Aeq} of 34 dB(A) at 4 h. About non-working days, the level decreases during the day and it is more variable (from 43 to 49 dB(A)). At night, the level is higher than during working days with a minimum of 37 dB(A) from 3 to 5 h.

Globally speaking, it is possible to see that the higher the typology of the road, the lower the noise level is, with almost non-existent overlaps among them. In all the types, noise levels are almost constant during the day and start to decrease around 22:00 h, being the quietest hour between 3:00 and 4:00 h. The level at 8:00 and 9:00 h (entrance to works and schools) is almost constant for the whole day and evening.

Therefore, according to the results of Cuenca, its day period embraces from 8:00 to 21:00 h and night period from 21:00 to 8:00 h. Furthermore, it can be stated that the more residential the street is, the higher the noise reduction during the night is. This behavior is due to the fact that in residential areas the traffic during the night is almost non-existent, whereas in other streets there is some more traffic during the night. This fact is also reinforced by the differences in the noise climate between day and night periods.

If working and non-working days are compared, it is possible to state that during the day the L_{Aeq} is generally lower during non-working days. At night the noise level in non-working days is increased. This increase in level is due to the increment in traffic because of leisure activities, which many times take place directly in the street.

Noise climate, shows a higher variability during the night period, being quite homogeneous for the day period. This fact, together with the higher noise levels during the day, indicates that there may be a significant number of similar noise sources that contribute to the noise pollution; meanwhile, during the night the number of noise sources is lower but more heterogeneous. Some differences also stand out if the road typology is considered, as those with a heavier traffic intensity (types 2–5) present values for the noise climate higher during the night period than during the day period; on the other hand, the trend is the opposite for the residential road types (6 and 7).

The singular noise events indicator (L_1) shows that in the surroundings of Cuenca there are no noise sources with emissions of high noise intensity within a short period of time (aircraft overflights, high-speed trains, booming sources, etc.). Nevertheless, it is stated that for the measurement points with lower L_{Aeq} and lower traffic intensity (for instance Acacia Uceta Street—type 7) there is a very significant variation of L_1 during the quietest hour (3:00–4:00 h); it means that in residential areas a single noise source can appear as extremely noisy, whereas for areas with more traffic intensity a single noise source is masked among all the noise sources.

Correlating the life in the city of Cuenca with the results, it can be set that the main cause of environmental noise is traffic [19]. The intensity of the traffic and its composition depends on the road typology. Taking into account the proposed road typology, the review of principal traffic noise models of [20] and the traffic intensity massively measured in Cuenca, a finite set of traffic intensity categories could be defined to refine, even more, the prediction model by assigning to each road not only its typology but also its traffic category.

4. Conclusions

In this paper, a road type classification has been developed taking into account road and urban characteristics. With this, seven types of roads have been defined: motorway, main distribution streets in the town, secondary distribution streets in the town, main distribution streets in a neighborhood, residential distribution streets, residential streets, and completely residential streets.

The long-time measurement results showed an almost constant level during the day, which decreases during the night. The more residential the road is, the more the level decrease during the night. Moreover, it can be observed higher level differences between day and night during working days than during non-working days.

This classification has been applied to the noise map of the World Heritage City of Cuenca. For that, a digital model was performed and adjusted with long time measurements carried out in roads classified inside different categories and with short-time measurements.

The methodology proposed allows allocating mean power and temporal behavior to each street from semi-permanent noise monitoring systems, and with this, the measurement cost of the noise map could be lowered, while maintaining a high level of accuracy.

This strategic tool may help to derive action plans to fight against the noise and reduce the number of people exposed to heavy noise.

Acknowledgements

This research has been possible thanks to project LIFE08-ENV/E/000110.

Conflict of interest

The authors declare that they have no known competing financial interests or personal relationships that could have appeared to influence the work reported in this paper.

Author details

José A. Ballesteros[1*], María Jesús Ballesteros[2], Samuel Quintana[1] and Marcos D. Fernandez[1]

1 Escuela Politécnica de Cuenca, Universidad de Castilla-La Mancha, Cuenca, Spain

2 Envirosuite, Madrid, Spain

*Address all correspondence to: josea.ballesteros@uclm.es

References

[1] European Commission. Directiva 2002/49/CE del Parlamento Europeo y del Consejo, de 25 de junio de 2002 sobre evaluación y gestión del ruido ambiental (Directive 2002/49/EC of the European Parliament and of the Council of 25 June 2002 relating to the assessment and management of environmental noise)

[2] BOE. Real Decreto 1513/2005 de 16 de diciembre, por el que se desarrolla la Ley 37/2003, de 17 de noviembre, del Ruido, en lo referente a la evaluación y gestión del ruido ambiental (Royal Decree 1513/2005 of December 16th to develop the Law 37/2003, of November 17th, about the noise regarding the assessment and management of the environmental noise)

[3] BOE. Real Decreto 1367/2007 de 19 de octubre, por el que se desarrolla la Ley 37/2003, de 17 de noviembre, del Ruido, en lo referente a zonificación acústica, objetivos de calidad y emisiones acústicas (Royal Decree 1367/2007 of October 19th to develop the Law 37/2003, of November 17th, about the noise regarding the acoustic zoning, quality objectives and acoustic emissions)

[4] E. C. W. G. A. of Exposure to Noise (WG-AEN). Good practice guide for strategic noise mapping and the production of associated data on noise exposure

[5] Brown A, Lam K. Urban noise surveys. Applied Acoustics. 1987;**20**(1):23-39

[6] Gómez V, Barrigón JM, Gozalo GR, Vaquero J, Sierra JM, Vílchez-Gómez R, et al. Acoustical environment of the medieval centre of Cáceres (Spain). Applied Acoustics. 2012;**73**(6):673-685

[7] Barrigón JM, Gómez V, Gutiérrez P, Méndez J. Study of Vitoria-Gasteiz by categorizing roadway characteristics. In: Proceedings of Internoise, Nice, France. 2000. p. 292

[8] Barrigón JM, Gómez V, Gutiérrez P, Alejandre L. Study of medium-sized cities of Extremadura by categorizing roadway characteristics. In: Proceedings of the 7th International Congress on Sound and Vibration (ICSV7), Garmisch-Partenkirchen, Germany. 2000. pp. 2413-2420

[9] Barrigón JM, Gómez V, Méndez JA, Vílchez-Gómez R, Trujillo J. An environmental noise study in the city of Cáceres (Spain). Applied Acoustics. 2002;**63**(10):1061-1070

[10] Gómez V, Barrigón JM, Méndez J, Vílchez-Gómez R, Vaquero J. Noise study of Salamanca (Spain) by means of a categorization method. In: Proceedings of Forum Acusticum, Seville, Spain. 2002. pp. 84-87985

[11] Barrigón JM, Gómez V, Méndez JA, Vílchez-Gómez R, Vaquero JM, Trujillo J. A categorization method applied to the study of urban road traffic noise. The Journal of the Acoustical Society of America. 2005;**117**(5):2844-2852

[12] Barrigón JM, Gómez V, Gozalo GR, Vílchez-Gómez R. Possible relation of noise levels in streets to the population of the municipalities in which they are located. The Journal of the Acoustical Society of America. 2010;**128**(2):EL86-EL92

[13] Bastián-Monarca N, Suárez E, Arenas JP. Assessment of methods for simplified traffic noise mapping of small

cities: Casework of the city of Valdivia (Chile). Science of the Total Environment. 2016;**550**:439-448

[14] Barrigón JM, Montes D, Gómez V, Gozalo GR, Vílchez-Gómez R. A proposal for producing calculated noise mapping defining the sound power levels of roads by street stratification. Environmental Pollution. 2021;**270**: 116080

[15] INE. National Institute of Statistics (Instituto Nacional de Estadística) [Internet]. 2018. Available from: http:// www.ine.es/ [Accessed: February 18, 2022]

[16] De Coensel B, Brown AL, Tomerini D. A road traffic noise pattern simulation model that includes distributions of vehicle sound power levels. Applied Acoustics. 2016;**111**: 170-178

[17] De Coensel B, Sun K, Wei W, van Renterghem T, Sineau M, et al. Dynamic noise mapping based on fixed and mobile sound measurements. In: EURONOISE 2015, Maastricht, France. 2015. p. 6

[18] ISO 1996-1:2016. Acoustics— Description, measurement and assessment of environmental noise— Part 1: Basic quantities and assessment procedures

[19] SICA. Information System about Noise Pollution (Sistema de Información sobre Contaminación Acústica) [Internet]. 2022. Available from: https:// sicaweb.cedex.es [Accessed: February 22, 2022]

[20] Garg N, Maji S. A critical review of principal traffic noise models: Strategies and implications. Environmental Impact Assessment Review. 2014;**46**:68-81. DOI: 10.1016/j.eiar.2014.02.001